The Beginner's Guide to Engineering:
Computer Engineering

quantum scientific publishing

The Beginner's Guide to Engineering:
Computer Engineering

James Lance

quantum scientific publishing

The Beginner's Guide to Engineering: Computer Engineering

ISBN-13: 978-1492981541
ISBN-10: 1492981540

Published by quantum scientific publishing

Pittsburgh, PA | Copyright © 2013

Cover design by Scott Sheariss

QUANTUM
SCIENTIFIC
PUBLISHING

Table of Contents

Unit Three

Appendix

Unit One

Section 1.1 – History of Computers

Section Objective

- Explain the history of computers

The Early Concept of a Computer

The history of computers and computer technology dates as far back as the 1200s. It was during this time when a process was created to solve problems with a written series of procedures. Skipping forward to the first primitive computing machines we find large, complicated, bulky and often times unpredictable computers. Rows of vacuum tubes and wires made these machines very complicated. It often took several rooms to fit one computer in to. Technology, research, and advancements have occurred within the world of computers which has resulted in a significant increase in power as well as a decrease in size. These advancements have also led to the significant decrease in the cost of computers.

Mechanical calculators were manufactured for sale as early as the 1640s. **Blaise Pascal** has been credited with inventing the first commercial calculator, a hand powered adding machine. Pascal was a French mathematician, physicist, and religious philosopher who contributed a lifetime of philosophies directed towards mathematics. Shortly after the first mass-produced calculator, a programmable machine was worked on by **Charles Babbage**. Charles Babbage invented the concept of the programmable computer. He worked much of his life towards computer designs and was eventually credited with inventing the first mechanical computer which led to more complex designs. His programmable machines were mainly created by using a set of gears and shafts. They were capable of translating short written work and simple calculus problems with moderate accuracy.

Calculator created by Blaise Pascal

Development by John Mauchly and J. Presper Eckert began on the Electronic Numerical Integrator And Computer (ENIAC) in 1943. They completed their work on the ENIAC in 1946. The US patent office recognized the **ENIAC** as the first computer. This machine quickly fell behind newer, more efficient technology after only three years. Rather than discontinue their same path, they decide to continue while working on a more modern machine, the EDVAC. They felt that they needed to make the machine appear more impressive to reporters than it may have actually been. In doing so, a team member added translucent spheres over the lights. These spheres were simply halved ping pong balls.

Detail of the back of a section of ENIAC, showing vacuum tubes

In 1944, the Havard Mark I was introduced. The Mark I was used to compute complex tables for the U.S. Navy. It used a paper tape to store instructions. IBM played a critical part in the machine's development. Early in 1945, with the Mark I stopped for repairs, a moth was found inside a piece of the computer which resulted in technical problems. Fixing the system became known as "debugging" from that day forward.

Computers begin to gain the confidence of a wider base and started to spread into different industries. The television station, CBS, used one of the 46 UNIVAC computers produced to predict the outcome of the 1952 Presidential Election. They delayed the prediction of the election and decided not to air the results for 3 hours because they did not trust the machine.

In 1961 **Fairchild Semiconductor International, Inc.** introduced the integrated circuit. A **circuit** is an electrical device that provides a path for electrical current to flow. Within ten years all computers use these instead of the transistor. The use of the circuit enabled the size of computers to go from building-sized to room sized. They also became much more powerful. The following year the Atlas became operational. The Atlas displayed many of the features that we see in today's systems such as virtual memory and instruction execution.

1969 became an important year towards the evolution of computers. Bell Labs developed their own operating system known as UNIX. ARPANet was launched which became one of the precursors to today's Internet. The personal computer (PC) was proposed for the first time by Alan Keys. Alan later became a designer for Apple. Intel was formed in 1969 after a group of technicians became unhappy with Fairchild Semiconductor and decided to form their own company. In 1971, Texas Instruments introduced the first pocket calculator which weighed 2.5 pounds.

Altair became the first personal computer and was marketed in kit form in 1975. The Altair featured 256 bytes of memory. Bill Gates helped write a BASIC compiler for the machine. Bill Gates later went on to form Microsoft which helped kick start the revolution of the personal computer. The next year Apple began to market PC's, also in kit form. It included a monitor and keyboard. In 1976, Queen Elizabeth went on-line with the first royal email message.

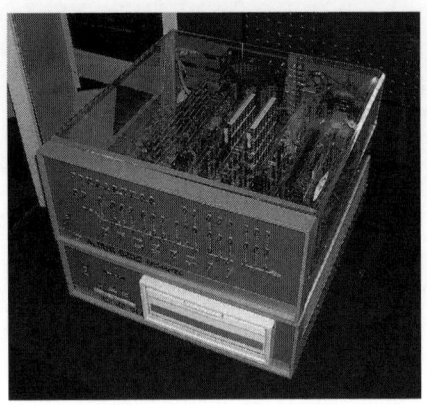

January 1975 Popular Electronics with the Altair 8800 computer

America would see an explosion of the personal computer onto the scene over the next few years. Microsoft, Apple and many smaller PC related companies successfully formed, while many more failed. By 1977 stores begin to sell PC's. Time magazine selected the computer as its Man of the Year in 1982. Companies today continue to strive to reduce the size and price of the personal computer while increasing the overall capacity.

Summary

The history of computers dates back to the 1200s. ENIAC became the first computer recognized by the US patent office in 1946. The use of circuits enables the size of computers to decrease and the power to increase. The first operating system, UNIX, was developed in 1969 while the first personal computer was proposed. The 70s see companies such as Microsoft and Apple successfully form which leads to further advancements and mass production of personal computers.

Concept Reinforcement

1. Discuss the significant importance of the breakthroughs during 1969.

2. What is the name of the first personal computer?

3. How did the term "debugging" start?

Section 1.2 – Basic Parts of a Computer

Section Objective

- List the basic parts of a computer

Computers come in all shapes and sizes. Portable computers and desktop computers are generally the primary sizes of computers for home, school, and office use. The portable computer comes in various sizes. These computers are typically referred to as laptops, notebooks, and hand-held computers. These generally represent different sizes. The laptop is typically the largest while the hand-held is usually considered the smallest.

The Foundation of a Computer

Regardless of size, each computer is comprised of many of the same main parts. To understand how a computer works on the inside we must first understand the CPU, memory and motherboard. The **central processing unit** (CPU) can be known as the "brain" of the computer system. The CPU is often called a processor. It is a chip whose speed determines how fast the personal computer (PC) operates. Everything that a computer does is managed by the CPU. It performs commands and instructions while controlling the operation of the computer.

Central Processing Unit

Memory is very fast storage used to hold data. Memory works directly with the CPU. There are different types of memory in a computer. Random-access memory (**RAM**) is used to temporarily store information which the computer is currently working. When you turn your computer off you will lose all data stored in RAM. Read-only memory (**ROM**) is a more permanent type of memory storage used by the computer for important data that does not change. Your computer is powered on and off each time by ROM.

All internal components of the computer connect to a main circuit board known as the **motherboard**. The CPU and memory as well as all other internal components are typically found here. Most of the internal components hook into the motherboard either directly or through cables. Many motherboards also include sound and graphic cards allowing you to listen, record and translate sounds or graphics.

Motherboard

Another component found within each computer is a real-time clock. By referring to the clock within your computer system all components in a computer can synchronize properly. Fans, heat sinks and cooling systems are also found internally and are used to regulate your internal computer temperature and allow performance to continue as expected.

Input / Output Devices

Every PC needs the ability to store information and interact with the user. Programs and documents need to be stored within the computer to allow the user to interact. The hard disk is large-capacity, permanent storage which is used to hold this type of information. The operating system is the basic software necessary to allow the user to interface with the computer. The operating system is stored within the hard disk.

Input/output devices, ports and networking are three different ways a typical computer interacts and communicates with the world around it. Even with the most powerful, fast, efficient internal components, there must be a way for the user to interact with them. This interaction is made available through input/output devices. The most common types of input / output devices in PCs include the monitor, keyboard, mouse, and removable storage.

The monitor, keyboard, and mouse are equally important towards viewing and manipulating data. The monitor is used as the primary device for displaying information from the computer. The keyboard is the primary device for entering information in the computer. The mouse is the primary device for navigating and interacting with the computer.

Storage and Communication

Removable storage devices allow you to add new information to your computer very easily, as well as save information that you want to carry to a different location. CD-ROM has become a widely used option for storage. It is used often for distribution of commercial software. CD-R (recordable) and CD-RW (rewritable), which can also record, are additional options. CD-RW discs can be erased and rewritten many times. DVD-ROM is similar to CD-ROM but has the capability of holding much more information. Flash memory is another type of removable storage. It provides fast, permanent storage which is based on a type of ROM.

Advancements in the different options a computer provides have justified the need for computers to allow you to connect to a wide selection of peripherals. Peripherals include your digital-video devices, such as camcorders or digital cameras. These peripherals connect to your computer through ports. A port is an interface that allows a computer to communicate with external equipment. One of the most common ports found on newer computers is the Universal Serial Bus (USB) which offers the ability to connect devices easily. Wireless connections, such as Bluetooth or Wi-Fi, allow you to communicate without connecting directly to your computer.

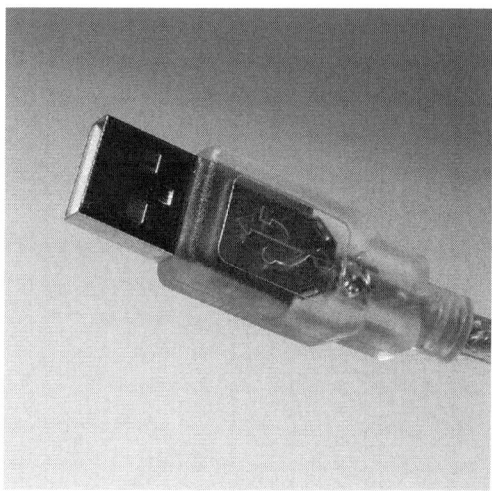

A USB plug

Summary

Portable and desktop computers come in all shapes and sizes. Each includes a CPU, memory, and motherboard which allow the computer to communicate together and perform at your desired speed. Each computer needs the capability to store information and interact with the user. The hard disk and operating system will enable these actions to occur. Computers have advanced with age and are now able to communicate with external devices through various input / output devices and ports.

Concept Reinforcement

1. What is a CPU?

2. What is the purpose of a clock within your computer?

3. Explain the difference between RAM and ROM.

4. List three types of input/output devices.

Section 1.3 – Information and Signals

Section Objective

- Define information and signals and describe how information is quantified

Information has long been difficult to specifically define and equate to computer theory and signal transmission. Scientists and technical engineers agree that information exists. They agree that information does not need to be perceived to exist nor does it need to be understood or interpreted to exist. However, when discussing information flow to and from a computer, we must compile some sort of understanding or definition together to justify how information is transferred and interpreted.

Theory of Information

Information can be represented in a number of diverse areas such as electronic signal flows, human brain functions, and the genetic code of DNA. **Claude Shannon** has been credited as the founder of information theory. In the late 1930s he determined that the principles of logic can be used to describe the two states (on and off) of electromechanical relay switches. Shannon believed that electrical circuits within computers, as we know them today, could complete operations of thought. Shannon developed a notion of information theory that information can be defined as the decision between two separate devices. As a result, the basic unit of information is designated as the bit, 1 or 0. Shannon became the first to identify and define information. His theory allows the concept of information to be mathematically traceable.

Shannon's theory of information now allows us to quantify information in a mathematical equation. In order to quantify information it is often paired with a signal. A **signal** is any electrical, light pulse or frequency whether in a wire, fiber or wireless medium. Information is usually transferred in signals or pulses. The term is somewhat generic and may refer to virtually anything that is generated and transmitted such as power, data, or control signals.

View of the lighthouse tower with the red light signal. On the right on the tower platform the electrical fog signal (fog horn).

How Information Can Be Quantified

Electrical signals can carry power and information. Signal processing theory incorporates the structure of the signals. It also identifies how systems affect signal components. Shannon's work describes how to efficiently represent signals whether they represent information or not. It also describes how communication channels can reliably communicate digital signals. It is tough to completely quantify how well a signal codes information because it is not certain exactly what form information takes. We do know, however, that information can be encoded into a signal.

An **encoder** is a device used to change a signal, data, or information into a code. This code may serve a variety of purposes. It may compress information to be stored or encrypted. It may also be used to translate information from one code to another.

An encoder is also used to compress information into a smaller unit, allowing additional information to be stored, encrypted, or processed. Compression is useful because it helps reduce the consumption of expensive resources, such as hard disk space. On the downside, compressed data must be decompressed to be used. For instance, a compression scheme for video may require expensive hardware for the video to be decompressed fast enough to be viewed as it is being decompressed. A good example of data compression is found within ZIP file format. ZIP format stores many source files in a single destination. Compressed information only works when both the sender and receiver of the information understand what is being encoded.

Shannon's work has developed into more advanced concepts such as digital signal processing (DSP). **DSP** is a process of identifying signals by assigning a specific sequence of numbers or symbols as well as the processing of these signals. Digital signal processing and analog signal processing make audio and speech signal processing, sonar and radar signal processing, and digital image processing possible.

Summary

Information has long been hard to define in reference to computer information. Claude Shannon has determined that a mathematical equation can be linked to information and held within signals. These signals allow information to travel to and from computers by the use of encoders and compression.

Concept Reinforcement

1. Explain Claude Shannon's impact in the world of computer information.

2. What is an encoder?

3. What is DSP?

Section 1.4 – Digital and Analog Information

Section Objective

- Explain the difference between digital and analog information

Analog vs. Digital

There are two types of telecommunication transmission, analog transmission and digital transmission. The word **digital** describes any system transmission based on discontinuous data or events. The important word here is discontinuous. Discontinuous transmission has breaks. Analog is the opposite of digital. **Analog** processes information in a continuous stream.

Analog transmission uses signals that are exact replicas of a sound wave or picture being transmitted. These signals are processed in a continuous wave. These waves include signals of varying frequency or amplitude to produce a continuous electric wave.

A telephone is a good example to use when discussing analog transmission. When a conversation takes place on a telephone an electronic current is transmitted though a wire and into the telephone receiver. To get from one end to the other the sound from the voice is reproduced in a specific, continuous pattern. Once this is completed, they are then converted back into sound waves.

Understanding the difference between digital and analog helps us to make sense of why digital computers are used more often than analog computers. That's not to say that all digital computers are better than analog. An analog computer could be more beneficial when built for specific outcomes such as processing images. Optical analog computers can be purchased that compute an image in as little time as it takes the image to pass through a special lens. This type of analog computer does its specific job very well, but is not able to do any other. On the other hand, a digital computer can be programmed to play games, spell-check documents, or tell time.

A 1960 Newmark analog computer used to solve differential equations

Digital computers are able to execute a wide range of jobs by translating information. Letters can be encoded by replacing every letter in the alphabet with its numerical position (1-26). A series of numbers can be associated with pitch and volume to encode different sounds within a computer. Similarly, an image can be encoded as a sequence of numbers that represent color and brightness. Each portion of the picture is measured. The computer then takes these numbers and converts them back into sounds, images, numbers or letters.

Communicating Between Computers

What happens when an analog signal is received by a digital computer? This signal must be converted to manipulate the analog signal into a digital signal. A **modem** is a device that is used to convert between analog and digital signals. This allows computers to communicate with each other across telephone lines. One computer sends a digital signal. The modem receives the digital signal and converts it to an analog signal. This allows the signal to be sent through a telephone line. When the signal reaches its destination another modem reconstructs the original digital signal so the computer can then process the original data.

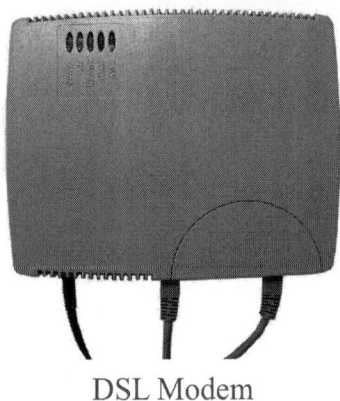

DSL Modem

Summary

Analog and digital are the two types of telecommunication transmissions. Analog processes information in a continuous stream while digital is discontinuous. An analog computer could be more beneficial than a digital when built for one specific outcome. However, a digital computer is capable of producing a wide range of outcomes including sounds, images, numbers and letters. A modem is used to convert between analog and digital signals.

Concept Reinforcement

1. Describe analog information.

2. Describe digital information.

3. What is used to convert an analog signal into a digital signal?

Section 1.5 – Bits and Binary Code

Section Objective

- Define bits and how they are used to create binary code

Since 2000 BC, humans have counted using 10 digits (0, 1, 2, 3, 4, 5, 6, 7, 8, 9). This is called "decimal base" or **base 10**. In the late 1930s, **Claude Shannon** determined that it was possible to carry out logical operations by using switches. If the switch were closed it equated to "true" and if a switch were opened it equated to "false." He then assigned the number 1 to "true" and 0 to "false."

Early example of humans using base 10 numeric system

This information encoding system that Claude Shannon used is now called binary. **Binary** uses two states which are represented by the digits 1 and 0 to encode information. This is what allows a computer to run.

Microprocessors are programmed with and operate in binary code. The values from all inputs form sensors and switches that are converted to binary form before they are read by the microprocessor. All memory operations, all data communication, and all output commands are performed with binary code.

The term **bit** means "binary digit", meaning 0 or 1 in binary numbering. The word bit is a shortening of the words *B*inary dig*IT*. It is the smallest unit of information which can be manipulated by a digital machine. Digital computers calculate by manipulating binary digits. Because bits are so simple to handle they can be made to stand for almost anything. This can include simple written words or different variables, codes, or instructions. Bits can be used to symbolize more complicated items as with logic or physics or even numerical calculations. Bits can even be used in recording images and sounds. When using a digital computer to solve or produce any of the given examples, certain bits inside the computer are arranged to represent what is being calculated or manipulated. The computer is able to apply different rules and variables just as a person would, only faster.

Base 10

To better understand the binary number system, the number system we use every day must be reviewed first. As mentioned previously, our number system is a base 10 number system. This means there are 10 different digits that we can use to make any number we want. The numbers are 0,1,2,3,4,5,6,7,8 and 9. Each numerical position in the base 10 number system has a value that is ten times the value of the previous position. When we view 1111111 as a binary number in the base 10 system the numeric 1 furthest to the right would represent 1 ($1 \times 10^0 = 1$). Continuing to move from right to left, the next numeric 1 would represent 10 ($1 \times 10^1 = 10$). The third numeric 1 from the right would represent 100 ($1 \times 10^2 = 100$) while the fourth would represent 1000 ($1 \times 10^3 = 1000$). This continues until you reach the numeric 1 on the far left which represents 10,000,000. The total value of 1111111 in base 10 is 11,111,111.

We can look at bits in the base 10 system in a different way as well. We can compare them directly to digits. We know digits are used in a single position to hold numerical value. For example, the number 3,489 has four digits. We understand that within the number 3,489, the 9 is holding the "1s place," while the 8 is holding the "10s place," the 4 is holding the "100s place," and the 3 is holding the "1,000s place."

$(3 \times 1000) + (4 \times 100) + (8 \times 10) + (9 \times 1) = 3000 + 400 + 80 + 9 = 3,489$

Another way to express it would be to use powers of 10.

$(3 \times 10^3) + (4 \times 10^2) + (8 \times 10^1) + (9 \times 10^0) = 3000 + 400 + 80 + 9 = 3,489$

This example shows each digit as a placeholder for the next higher power of 10, starting in the first digit with 10 raised to the power of zero.

Base 2

Computers use the **base 2** number system. This means there are 2 different digits called bits that we can use to make any number we want rather than using digits. Where digits have 10 possible values ranging from 0 to 9, bits only have two possible values: 0 and 1. Eight bits make a byte. One million bytes makes a mega-byte. One billion bytes makes a giga-byte. Bits are rarely seen alone in computers. They are almost always bundled together into 8-bit collections, and these collections are called bytes. With 8 bits in a byte, you can represent 256 values ranging from 0 to 255.

Each numerical position in the base 2 number system has a value that is two times the value of the previous position. Multiple binary bits can be combined to form any number possible. When we view 1111111 as a binary number in the base 2 system the numeric 1 furthest to the right would represent 1 ($1 \times 2^0 = 1$). Continuing to move from right to left, the next numeric 1 would represent 2 ($1 \times 2^1 = 2$). The third numeric 1 from the right would represent 4 ($1 \times 2^2 = 4$) while the fourth would represent 8 ($1 \times 2^3 = 8$). This continues until you reach the numeric 1 on the far left which represents 128. The total value of binary number 1111111 in base 2 is 255.

Remembering that a binary number is composed of only 0s and 1s, how do we figure out what the value of the binary number 1101 would be by using the base 2 system? You do it in the same way we did it above for the number 3,489, but you use a base of 2 instead of a base of 10.

$(1 \times 2^3) + (1 \times 2^2) + (0 \times 2^1) + (1 \times 2^0) = 8 + 4 + 0 + 1 = 13$

It is easy to see the value in the binary system when looking at the progress made with audio and video information. Audio and video information was first translated into computer data in the 1960s. Every point in a color video image and every instant of sound were translated into 1s and 0s. This concept allows a television series or movie to be saved digitally in a computer. Unfortunately this was virtually impossible since a digital television signal took so many bits per second to be encoded. In the late 1980's, however, digital compression allowed the pictures to be transmitted in a highly abbreviated form, making the encoding of these signals to a computer possible.

Summary

Binary is an encoding system that allows computers to run. Binary uses two states which are represented by digits 1 and 0 to encode information. These digits are called bits. A bit is the smallest unit of information which can be manipulated by a digital machine.

Concept Reinforcement

1. Explain the difference between the base 10 and base 2 numbering systems.

2. What is a bit?

3. Explain how binary code is used within computers.

Section 1.6 – Using Protocols to Organize Information

Section Objective

• Explain using protocols to organize information

Computers communicate in a way in which they must have everything defined and structured. They must know in advance how much and in what format information is to be exchanged. To accomplish this standard of communication, methods of information transfer and processing have been devised. These standards are referred to as protocols. **Protocols** ensure that computers everywhere can talk to one another.

There are a variety of standard protocols from which programmers can choose. Each has particular advantages and disadvantages. Some protocols are simpler than others, some are more reliable, and some are faster. Protocols are used to determine the type of error checking to be used and data compression method, if any. They also determine how a device will indicate that it has sent a message and how the receiving device will indicate that the message has been received.

OSI Model

Rules about appearance, speaking, listening and understanding are considered human communication protocols. Each of these rules represents different layers of communication. People rely on these rules for effective communication. While layers of human communication are typically not as defined or formal as computer communication, protocol still exists. The need for protocols also applies to network devices. Computers have no way of learning protocols so network engineers have written rules for communication that must be strictly followed for successful communication. These rules have been formalized into what is known and accepted as the **Open Systems Interconnection Reference Model** (OSI Reference Model or OSI Model).

Protocols are organized within the OSI model according to the level of detail required for information transmission. Protocols at the lower levels of the model (shown toward the bottom) focus on bit transmission. Higher level protocols focus on how bits are organized to represent information, what kind of information is defined by bit sequences, what software needs the information, and how the information is to be interpreted.

Each layer includes similar functions that provide services to the layer above it and receives service from the layer below it. The OSI model divides network architecture into seven layers. These layers, from top to bottom, are the Application, Presentation, Session, Transport, Network, Data-Link, and Physical Layers.

Layer 7: Application Layer

The application layer is the OSI layer closest to the end user, which means that both the OSI application layer and the user interact directly with the software application. Application layer functions typically include identifying communication partners, determining resource availability and synchronizing communication.

Layer 6: Presentation Layer

The presentation layer works to transform data into the form that the application layer can accept. This layer formats and encrypts data to avoid compatibility problems prior to being sent across a network.

Layer 5: Session Layer

The Session Layer controls the dialogues/connections (sessions) between computers. It establishes, manages and terminates the connections between the local and remote application.

Layer 4: Transport Layer

The Transport Layer provides transparent transfer of data between end users, providing reliable data transfer services to the upper layers. The Transport Layer controls the reliability of a given link through flow control, segmentation, and error control. The Transport Layer can keep track of the segments and retransmit any that fail.

Layer 3: Network Layer

The Network Layer performs network routing functions. It may also report delivery errors. Routers operate at this layer which send data throughout the extended network and make the Internet possible.

Layer 2: Data Link Layer

The Data Link Layer provides the functional and procedural ability to transfer data between network entities and to detect and possibly correct errors that may occur in the Physical Layer.

Layer 1: Physical Layer

The Physical Layer defines the electrical and physical specifications for devices. It defines the relationship between a device and a physical medium. The Physical Layer will tell one device how to transmit to the medium, and another device how to receive from it.

Summary

Computers communicate in a way in which they must have everything defined and structured. To accomplish this standard of communication, methods of information transfer and processing have been devised. These standards are referred to as protocols. Protocols ensure that computers everywhere can talk to one another. There are a variety of standard protocols from which programmers can choose. Computers have no way of learning protocols so network engineers have written rules for communication that must be strictly followed for successful communication. These rules have been formalized into what is known and accepted as the Open Systems Interconnection Reference Model (OSI Reference Model or OSI Model). The OSI model divides network architecture into seven layers. These layers, from top to bottom, are the Application, Presentation, Session, Transport, Network, Data-Link, and Physical Layers.

Concept Reinforcement

1. List and define the layers of the OSI Model.

2. Explain what a protocol is and describe its uses?

Section 1.7 – Protocols for Sending Data

Section Objective

• List and describe protocols for sending data

File Transfer Protocol

There are many different protocols used for a variety of specific scenarios. FTP can be used when sending or copying data over a network from one computer to another. FTP stands for **File Transfer Protocol**. The main purpose of FTP is to transfer files over a network from one computer to another without having to physically copy the data onto a disk or physically install it onto another computer.

You can use FTP to move files such as word processing files, HTML pages from the Internet or even simple text files between two computers. It allows a user to both retrieve information and deposit information on a server. The receiving of information is called **downloading** while the giving of information is called **uploading**.

FTP was invented in 1985 as a clear way of passing information from point A to point B. While FTP has been used less as the Internet has come along, it still remains a useful protocol. FTP revolutionized data exchange and has been credited with much of the growth of the Internet because it so greatly streamlined file swapping.

FTP allows files to be sent to anywhere in the world. It doesn't matter what type of computer you use, what type of operating system it runs off of, or even how you are connected to the internet. Making a phone call is a simple example that can be used to relate to FTP. You can pick up your phone whenever you would like and make a call to a friend in Europe. As long as your friend has a phone the call will always go through, regardless of the type of phone or the long distance carrier. In much of the same fashion, all computers carry a type of FTP program which allows communication to occur between any two networked computers.

Every operating system (OS) has its own way of organizing files, which are arranged in a way that makes it easiest for each OS to access data. Without FTP it would be difficult to communicate between two computers using different operating systems. For example, you could be working from a Windows operating system and the computer you want to retrieve files from is running Unix. Your Windows machine has no idea where to find the file you requested because it doesn't use the same logic as the Unix machine. FTP allows these different operating systems to communicate.

FTP lets the two computers share a few crucial words of one language so they look the same to each other. These words don't necessarily change either computer because the protocol is just simple enough that both systems can understand the language the other is using. The basis of this critical language follows a list of simple commands. The most popular of these commands include OPEN, DIR, GET, SEND, and CLOSE.

OPEN: This command initiates a connection between your computer (the client) and the other computer (the server) so that files may be exchanged.

DIR: This lets your machine request a listing of the directories and their contents on the remote host.

GET: This command requests that the file be transferred from the remote to the local host.

SEND: This command works in reverse, delivering a file from your computer to a remote one.

CLOSE: This ends the file transfer session.

Every computer knows and understands those commands. No matter what system it uses for storing and organizing files, it processes these requests which allow it to send a file from one computer to another.

Summary

FTP stands for File Transfer Protocol. The main purpose of FTP is to transfer files over a network from one computer to another without having to physically copy the data onto a disk or physically install it onto another computer. You can use FTP to move files such as word processing files, HTML pages from the Internet or even simple text files between two computers. FTP lets two computers share a few crucial words of one language so they look the same to each other. The most popular of these commands include OPEN, DIR, GET, SEND, and CLOSE.

Concept Reinforcement

1. What does FTP stand for?

2. What is the difference between downloading and uploading?

3. Explain how FTP makes communication between two computers easier.

4. List and describe 5 basic commands recognized by all computers.

Section 1.8 – Word Processor and Internet Protocols

Section Objective

- Explain word processor and Internet protocols

Computers can communicate different types of information such as text, pictures, voice, and more. Each of these is communicated through a protocol known as TCP/IP. TCP/IP stands for **Transmission Control Protocol / Internet Protocol** and is made of two sets of rules for communication over the Internet.

Transmission Control Protocol (TCP) is responsible for dissecting pieces of information so they can be transferred over the Internet and then reassembled. **Internet Protocol (IP)** manages addresses and guides files to the intended destination. Combined, TCP/IP is simple and efficient.

TCP/IP works much like postal delivery. Delivery must be managed correctly in order to be delivered accurately to the correct address.

TCP / IP

Transmission Control Protocol and Internet Protocol are two different protocols that are often found linked together. To find protocols linked together is not uncommon. In the case of TCP and IP the functions complement each other to allow a task to be completed more efficiently. TCP/IP is what carries out the basic operations of the Web. It is also used on many local area networks.

When information is sent over the Internet, it is generally broken up into smaller pieces or packets. Transmission speed is increased by the use of packets. The speed is increased because of the efficiency that packets allow. Different parts of a message can be sent by different routes if the path seems more efficient. These packets eventually come together and are reassembled at the final destination. Sending information in packets also minimizes the opportunity of losing all your data in the transmission process. TCP is the means for creating the packets, putting them back together in the correct order at the end, and checking to make sure that no packets got lost in transmission. If necessary, TCP will request that a packet be resent.

Routing information to the proper address is the responsibility of IP. Every computer on the Internet has an IP address. The IP address is unique to each computer, much like the address of your home. Every packet sent will contain an IP address showing where it is supposed to go. A packet may go through a number of computer routers before arriving at its final destination. IP controls the process of getting everything to the designated computer.

Mail Protocols

There are a variety of protocols for sending and receiving email. The most common protocol for sending mail is **Simple Mail Transfer Protocol** (SMTP). An Internet address for a SMTP server must be entered when configuring email clients such as Outlook Express. **Post Office Protocol** (POP) is the most common protocol for receiving mail. Email clients such as Outlook Express require an address for a POP server before they can read mail. The SMTP and POP servers may or may not be the same address. Both SMTP and POP use TCP for managing the transmission and delivery of mail across the Internet.

A more powerful but less common protocol for reading mail is **Interactive Mail Access Protocol** (IMAP). IMAP is more common in business environments as it allows for the reading of individual mailboxes at a single account. IMAP also uses TCP to manage the actual transmission of mail.

Hypertext Transfer Protocol

Many of our personal email addresses are created through Web based mail such as Yahoo or Hotmail. Web mail involves the same protocol as a Web page. Web pages are constructed according to a standard method called **Hypertext Markup Language** (HTML). An HTML page is transmitted over the Web in a standard way and format known as **Hypertext Transfer Protocol** (HTTP). This protocol uses TCP/IP to manage the Web transmission.

A related protocol is **Hypertext Transfer Protocol over Secure Socket Layer** (HTTPS). A Web page using the HTTPS protocol will have https: at the front of its URL. This provides security for sensitive data as the transmission of information through HTTPS is encrypted.

Summary

Computers can communicate different types of information such as text, pictures, voice, and more. Each of these is communicated through a protocol known as TCP/IP. Transmission Control Protocol (TCP) is responsible for dissecting pieces of information so they can be transferred over the Internet and then reassembled. Internet Protocol **(IP)** manages addresses and guides files to the intended destination. Combined, TCP/IP is simple and efficient. The most common protocol for sending mail is Simple Mail Transfer Protocol (SMTP). Post Office Protocol (POP) is the most common protocol for receiving mail. An HTML page is transmitted over the Web in a standard way and format known as Hypertext Transfer Protocol (HTTP).

Concept Reinforcement

1. Describe what TCP stands for and explain how the protocol is used.

2. Describe what IP stands for and explain how the protocol is used.

3. Explain the benefit of two protocols being used together to form TCP/IP.

Section 1.9 – Saving Information

Section Objective

- Describe saving information

Storage Devices

When we think of saving information on or to our personal computer we typically think of devices such as USB sticks, zip drives, floppy disks or CDs. These are all considered types of storage, or **secondary storage**, where data can be saved. Secondary storage differs from **primary storage** in that it is not directly accessible by the CPU.

The computer usually uses its input/output channels to access secondary storage and transfers the desired data using primary storage. Secondary storage is non-volatile which means it does not lose the data when the device is powered down.

Off-line Storage

Off-line storage, also known as disconnected storage, is computer data storage on a device that is not under the control of a processing unit. Information is recorded, usually in a secondary storage device, and then physically removed or disconnected. The information must be inserted or connected by a human operator before a computer can access it again.

Off-line storage is used to save and transfer information. Off-line storage devices are made to be easily transported as their size and shape is typically very manageable. The benefit of off-line storage can be crucial. In the case of a disaster it is vital to have important information stored in a remote location where it will be unaffected. This allows for the data to be recovered by transferring it to an unaffected computer. Another benefit to off-line storage is its increase towards general information security. Since this information is physically inaccessible from a computer it is not susceptible to computer theft or virus.

Most secondary storage media for personal computers are also used for off-line storage. Optical discs, flash memory, and removable hard drives are popular. Older examples are floppy disks, Zip disks, or punched cards.

This picture shows an example of a dated storage media, the punch card.

Backup Today, Not Tomorrow

Everyone knows about losing data in a computer crash or as the result of a virus. Backing up, or saving, your data to a storage device external to your PC is one of the most powerful ways to protect against this. Backing up your computer is the process of making a secondary copy of information. This data is typically the type that cannot be replaced or that you know will be needed in the future. While backing up your computer will not save your computer from crashing, or keep viruses from being able to infect it, it will ensure that no matter what happens to your computer you will still have access to the files that are important to you.

ZIP drives, CD-Rs, DVD-Rs, external hard drives, among others, make it easy to backup your entire computer. ZIP drives even come with a special utility that is designed specifically for the purpose of backing up your entire computer onto ZIP disk. These formats can hold hundreds of megabytes and even more information on a single disk, and can prove very powerful in the backing up of your information.

This picture shows an example of an external hard drive which is an efficient storage device because of its physical size and ability to store a large amount of data.

Backup allows you to save your operating system as well as all the programs and utilities that you have saved to your computer over the years. Personal items such as pictures, music, or documentation may be irreplaceable if not saved to a secondary device. Restoring your entire computer from disk is fast, easy, and can include anything and everything that you had on your computer.

Backup is a powerful tool and everyone who uses a computer should have at least some of their files in backup. We all know that someday the potential of a disaster happening is possible. Saving your important data is very powerful and a task that certainly should be completed regularly.

Summary

Secondary storage differs from primary storage in that it is not directly accessible by the CPU. The computer usually uses its input/output channels to access secondary storage and transfers the desired data using primary storage. Secondary storage is non-volatile which means it does not lose the data when the device is powered down. Off-line storage, also known as disconnected storage, is computer data storage on a device that is not under the control of a processing unit. Information is recorded, usually in a secondary storage device, and then physically removed or disconnected. Backup is a powerful tool that should be used by everyone who owns a computer. We all know that someday a disaster could happen, and when that day comes you will be happy to have your files in backup.

Concept Reinforcement

1. Explain the difference between primary and secondary storage.

2. Define offline storage.

3. Discuss the importance of keeping data backed up.

Section 1.10 – Images

Section Objective

• Discuss the types of electronic images

A digital image is a representation of a two-dimensional image using ones and zeros (binary). Vector images and raster images are the two different types of basic image types.

Vector Image

A **vector image** takes a digital image and gives it a two-dimensional or three-dimensional shape. Vector uses a sequence of commands or mathematical statements to create the digital image. Vector graphics are saved in a sequence of statements. Rather than save each bit within a file, a vector graphic file will describe a series of points to be connected. This results in a much smaller file.

Storing images using this method is effective if they are drawings or diagrams. It would be very hard to have a photograph stored as a vector image. Current internet browsers cannot display vector images on their own.

Raster Image

A **raster image**, also called a bitmap, is another way to represent digital images. A raster image represents an image in a series of bits of information which translate into pixels on the screen. These pixels form points of color which create an overall finished image. A **pixel** is the smallest display element that makes up an image you see on a computer or television. A typical image contains millions of pixels which is why digital camera output is defined in megapixels (mega=millions). The resolution and picture quality increase as the pixel count increases.

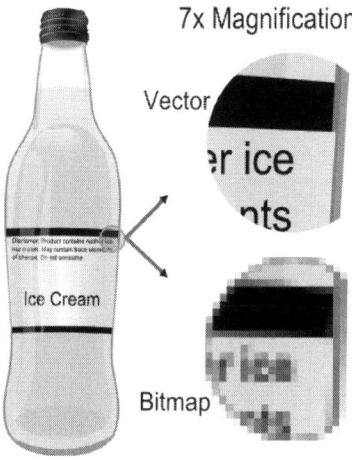

This image shows the difference between a vector image and a raster, or bitmap, image.

Compare the zoomed-in areas in raster and vector images in the picture above. Notice that the vector image is smooth while the raster image, or bitmap, shows visible squares, or pixels. Both images are magnified as the same level yet the vector image remains clear while the raster image begins to distort.

Converting Images

Converting a vector image to a raster image can be done with relative ease. By selecting the size of pixels an image can be converted. Pixel size is measured by length and width. This pixel value is called resolution and it is measured in dots per inch (DPI). The process of converting a vector image into a raster image graphic is called bitmap.

A vector file is sometimes called a geometric file because of its use of connecting images through mathematical equations. Most images that are created with tools such as Adobe Illustrator and CorelDraw are in the form of vector image files. Animation images are also usually created as vector files. Vector image files are easier to modify than raster image files.

Color Combinations

LCD or CRT is the most natural way to represent color to the human eye, although there are multiple options. A LCD computer monitor displays colors by combining various intensities of red, green, and blue. A color can be described by using three numbers representing intensities of red, green, and blue on a scale from 0 to 100%.

The accuracy of these numbers is called color depth of an image. **Color depth** determines the number of colors in an image. Values are given either by specifying the total number of displayable colors or in bits per pixel. If color depth is 24 bits per pixel (BPP), the number of displayable colors is 224, that is roughly 16 millions.

File Formats

In a raster image each pixel is described by specifying values for red, green, and blue color components. To store an image, three sets of numbers must be stored ranging from 0 to 100. Each number represents red, green, and blue. This can add up to a significant amount of data. Large amounts of data result in more space being taken up on your hard drive. This also increases the amount of time it will take to download to or from your computer. File formats are used to reduce the amount of data needed to be stored, downloaded or sent.

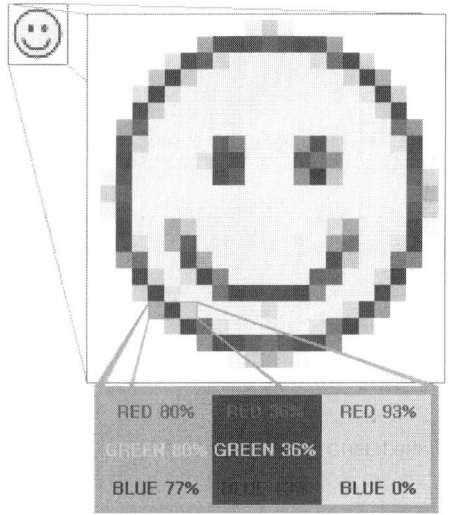

RED 80%   RED 98%   RED 93%
GREEN 80% GREEN 36% GREEN 2%
BLUE 77%   BLUE 12%   BLUE 0%

This raster image shows how each pixel can represent a different
color combination by adding different values for red, green, and blue.

File formatting includes data compression. **Data compression** is the process of storing data in a format that requires less space than usual. Data compression is useful in communications because it enables devices to transmit or store the same amount of data in fewer bits. It is inefficient to store pixels within an image individually. Because of this, file formats are used to determine how data will be reduced with a format change and if the image can be restored exactly back to its original version.

JPEG is a widely used file format. It was created to store photographs. JPEG files are very efficient at reducing the file size. However, when compressed to less than 10% of the original image size the original image may not be remembered exactly. This causes some details to be omitted. **PNG** is an acronym of Portable Network Graphics. This file format is intended for artificial images. These images have sharp edges or areas filled using a single-color. Last of the image file format used on internet is **GIF**. Its main disadvantage is the limited number of colors in image (256). Due to this limitation, GIF is not used often for photographs or large artificial images. GIF is best used for small animated images or when needing a transparent background.

When formatting impacts the original version of an image enhancement can be used. **Image enhancement** is the process of improving the quality of a digitally stored image by manipulating the image with software. Making an image lighter or darker or increasing or decreasing the contrast are all possible through image enhancement.

Summary

A digital image is a representation of a two-dimensional image using ones and zeros (binary). Vector images and raster images are the two different types of basic image types. A vector image takes a digital image and gives it a two-dimensional or three-dimensional shape by using a sequence of commands or mathematical statements to create the digital image. A raster image represents an image in a series of bits of information which translate into pixels on the screen. A pixel is the smallest display element that makes up an image you see on a computer or television.

Concept Reinforcement

1. Describe the difference between a vector image and a raster image.

2. Explain data compression and its uses.

3. List and explain three potential data image file formats.

4. Explain image enhancement.

Section 1.11 – Converting Images to Bits

Section Objective

- Explain the process of converting images to bits

Converting images to bits is a process used by computers to represent a picture in a format that can be easily interpreted and manipulated. Processing bits is how a computer works, regardless of the type of information. Images are read by computers in much of the same fashion. An image is often represented as a bitmap to make the conversion process from an image to bits possible. A **bitmap** is a representation of the entire image consisting of rows and columns filled with dots. These dots allow the computer to form a graphic image. This image can then be stored within the computer memory.

The value of each dot is stored in one or more bits of data. Each dot does not necessarily need to be filled. Neutral or colorless images may only need to be represented by one bit. However, colors and shades of gray require data consisting of more bits to better identify the color. The more bits used to represent a dot, the more colors and shades of gray that can be represented.

The density of the dots is known as the resolution. The **resolution** determines how sharply the image is represented. This is often expressed in dots per inch (dpi). DPI is the number of rows and columns within the bitmap. For example, the image may equate to 640 rows and 480 columns of dots within the bitmap. The dpi of this image is 640 by 480. Bitmaps are translated into pixels when an image is being displayed on a computer monitor or printed to a printer. The difference in total bits used can be tremendous based on the type of picture. For a two-color, black and white picture, a 100 by 100 pixel picture requires 100 x 100 or 10,000 bits. For a full-color picture, each pixel of the image requires 24 bits, 8 each for red, green, and blue. The 100 by 100 pixel picture now jumps in size to 240,000 bits.

This picture represents image output based on resolution.

Because images can be so different in size there are different options for storing an entire image rather than individually storing colors for each pixel. These options help reduce image size. Formats like GIF rely on identifying repeating patterns for compression. Formats like BMP (and GIF) rely on using an indexed color table approach. Instead of storing the specific color for each pixel with each pixel, the BMP format relies on an indexing approach. All the necessary colors are placed in a table, and an index into the table is stored for each pixel.

The indexed lookup table size determines how many bits are needed for each pixel. For a 16 color table, you need 4 bits. For 256 colors, you need 8 bits. This means that the same example that we used above for a 100 by 100 pixel image would now require 40,000 or 80,000 bits with the indexed approach, instead of the 240,000 bits with the true color approach.

Bitmap graphics are often referred to as raster graphics. The other method for representing images is known as vector graphics. Vector graphics defines each shape within an image with a mathematical formula. Vector graphics are more flexible than bit-mapped graphics because they look the same even when you scale them to different sizes. Bitmap graphics become ragged when you shrink or enlarge them.

Summary

Converting images to bits is a process used by computers to represent a picture in a format that can be easily interpreted and manipulated. An image is often represented as a bitmap to make the conversion process from an image to bits possible. A bitmap is a representation of the entire image consisting of rows and columns filled with dots. The more bits used to represent a dot, the more colors and shades of gray that can be represented. The density of the dots is known as the resolution which determines how sharply the image is represented.

Concept Reinforcement

1. Why is an image often represented as a bitmap?

2. Explain resolution.

3. Explain what a dot within a bitmap may consist of.

Section 1.12 – Synthesizing Images

Section Objective

- Describe the process of synthesizing images

A computer can take descriptions of the world and create pictures using any rules you choose. This process is called **image synthesis**. Image synthesis is often referred to a **rendering**. It is the process of using a computer program to generate an image from a model. Image synthesis turns the rules of geometry and physics into pictures that mean something to people. These rules often include viewpoint, texture, lighting, and shading information which have been used in architecture, video games, simulators, movie or even TV special effects and design visualization.

Spiral Sphere and Julia, Detail, a computer-generated image
created by visual artist Robert W. McGregor

Computer graphics is one of the newest forms of visual media. Image synthesis is a major sub-topic of 3D computer graphics. Image synthesis is one of the last major steps in the process of computer graphics, giving the final appearance to the models and animation. Image synthesis has come so far that it makes it tough to determine if an image is a photo, a model or a computer rendering. Just take a look at King Kong the movie. King Kong was fully virtual and digital, yet he was so life-like that he was able to draw strong emotions from the audience.

Image synthesis creates new images from some form of an image description. Test patterns, image noise, and computer graphics are each examples of images that are typically synthesized. Test patterns are scenes with simple two dimensional geometric shapes. Image noise are images with which contain random pixel values. Computer graphics are based on two or three dimensional geometric shape descriptions.

Image synthesis usually involves a digital image or raster graphics image. When rendering is used with three dimensional computer graphics the process can become long and tedious. This process is often referred to as **pre-rendering** or real time rendering. Pre- rendering is typically used when creating animated movies, while real time rendering is used for creating three dimensional video games.

Digital rendering takes advantage of a number of features that may be altered in a variety of ways in creating a final product. Shading is used to add brightness and color. Texture-mapping and bump-mapping allow small detail to be applied to the final image. Mirror-like images can be added through reflection. The transmission and detail of light is determined though techniques such as transparency, translucency, refraction, and fogging. Motion blur, shadows and depth of field add effects that give the appearance of movement or placement within a focused area.

Rendering programs have gone from limited, highly expensive graphic technologies to programs that can now be found at little to no cost through the Internet. A wide variety of rendering programs are now available. Depending on the projected outcome of the project, some programs will be found as a package to an animation or modeling suite. Other options are stand-alone packages. Each of these programs or software applications combines geometry and physics into pictures that mean something.

Summary

Image synthesis is often referred to a rendering. It is the process of using a computer program to generating an image from a model. Image synthesis creates new images from some form of an image description. Test patterns, image noise, and computer graphics are each examples of images that are typically synthesized. Digital rendering takes advantage of a number of features that may be altered in a variety of ways in creating a final product.

Concept Reinforcement

1. Explain image synthesis.

2. What type of images are most commonly synthesized?

3. Discuss the process of digital imaging.

Section 1.13 – Storing Images

Section Objective

- Explain how to store images

The world of converting, storing, and even creating images has been made much easier with the development of digital cameras. As we have learned, a digital image is represented by using 1s and 0s (binary). This makes these digital, two-dimensional images, much easier to convert directly into your computer. Once converted into your computer you must make a decision on where the images should be stored. While the conversion process is made relatively simple, there are a number of options you must consider in order to guarantee your images are safe and easily accessible. There are various options for image storage. Depending on the situation, it may be wise to consider more than one option when storing your digital images.

Memory Card

A memory card is a method used to store your images as you shoot with your digital camera. These memory cards can be inserted directly into your digital camera and have a set storage capacity which will determine how many pictures can be taken per card. The primary intention of these cards is for temporary storage until you are able to shift your images over to your computer. If you decide not to convert your images you can simply leave them on your memory card. This method of storing is an option for someone who does not take many pictures. The card will allow you to take pictures until it is filled. For most, this method is best used for temporary storage, as it was intended.

Sony Memory Stick can be used for some Sony digital cameras for storing images

Image Storing Options

After you have shifted the images over to your computer from your memory card, the images will then be located on your hard drive. The hard drive is where most people store the bulk of their images. This is a good option but should not be your only storage destination. If you have all of your images on your hard drive and it crashes or becomes infected with a virus, each of the images stored within your hard drive will be lost. The hard drive is an efficient destination for storing images because they are easily accessible within your computer. While you will want to use your hard drive for storage you may also want to consider a back up to all images that are transferred to your hard drive. These back up devices are known as **secondary methods**.

Printing images is a way of storing them as well. However, you will never want to use this as the only method if you have any possibility of needing the original image file. A printed image will need to be scanned in order to convert it back to a format which can be converted back to a computer. A scanned photo print will always have far less image quality than the original file. Printed images are also susceptible to potential damage from water, heat or light. Printing images is one way to create a backup, and is an easy way to quickly see what the images look like, but should not be your solution for storing an image in a secondary location.

One of the most common image backup systems is to use a compact disc (CD). Depending on your need, the CD can be a wonderful way to store images. CDs can be rather efficient and are relatively simple and inexpensive to use. You will, however, need a CD-burner to get your images from the computer to the CD if you use this method of storage. CDs are not free from risk. The biggest concern with storing images to a CD is the potential for a scratch. One scratch could potential damage every image on the disk. Also, while individual CDs are relatively inexpensive the storage space may not go as far as you may want per CD. To store all of your images may cause the CDs to add up. This makes a reliable system for organization a must. Dates, themes, or descriptions of the stored images should be written down with each CD. CDs may not be right for everyone or every situation but is still a good option for storing your images in a secondary location.

A DVD follows many of the benefits and concerns of a CD. The primary difference is storage space. A DVD will allow much more storage than a CD which may significantly cut down on the amount of discs you need to store and organize. However, the danger of a scratch is just as much of a concern with DVDs. These should be stored in a safe environment to help protect your images.

Online storage is another great storage option. Web sites have been created to allow you to download and store images for little to no cost. These images can easily be shared with friends or family. One drawback to consider with online storage is the limit to the image file size. If you are taking images that are too large the online site may not allow access for storage.

External USB hard drives are a great option to consider as a back up to your hard drive. This is considered by many to be the most efficient option to safely back up your images. They hold large amounts of data, have become less expensive to purchase, and tend to be more reliable than burning images to a CD or DVD. An external USB hard drive allows you to plug the drive directly to your computer, edit your images, and resave them since it acts as a regular drive. An external USB hard drive becomes a great option for storing large amounts of images in a safe place away from your computer.

We never hope for anything to go wrong with our computer. Backing up and storing images in a secondary location will put you at ease if something were to go wrong and your images within your hard drive were no longer accessible. Selecting two of the storage methods mentioned and updating them regularly should prevent your images from ever being completely lost. Remember to file them so they are easy to organize and find.

Summary

There are various options for image storage. A memory card is a method used to store your images as you shoot with your digital camera. The hard drive is where most people store the bulk of their images. One of the most common image backup systems is to use a compact disc (CD). Web sites have been created to allow you to download and store images for little to no cost. External USB hard drives are a great option to consider as a back up to your hard drive.

Concept Reinforcement

1. List three popular options for storing images.

2. Explain the purpose of a memory card.

3. What is the difference between a CD and a DVD?

Section 1.14 – Virtual Reality Modeling Language

Section Objective

- Describe Virtual Reality Modeling Language (VRML)

VRML stands for **Virtual Reality Modeling Language**. VRML allows Web developers to create three-dimensional (3-D) space and 3-D objects in full color. This also includes using special texture, animation, and lighting effects. This allows users to move in three dimensions on a VRML Web page in much of the same fashion as you would with a video game or flight simulator. VRML may be used in a variety of application areas such as engineering and scientific visualization, multimedia presentations, entertainment and educational titles, web pages, and shared virtual worlds.

Purpose of VRML

Virtual Reality Modeling Language was developed by the Web3D Consortium and was designed for use on the Internet. VRML was recognized as an international standard by the International Organization for Standardization (ISO) and the International Electrotechnical Commission (IEC) in December, 1997. While officially recognized in 1997, VRML had been widely used to share and publish data prior to that. It had primarily been used with CAD, animation, and 3D modeling programs. Each of these programs today typically has a utility that allows conversion to or from VRML.

Foundation of VRML

VRML can be used as a scene description language which has become a standard for the Web. VRML is used to create scenes. The language is used to describe the geometry and behavior of three-dimensional scenes and is designed to be used on the Internet, intranets, and local client systems. Each **scene** represents a 3D space or object. Scenes are combined to make **worlds**. VRML scenes can be distributed over the World-Wide Web and browsed with special VRML browsers. Scenes can be easily connected with other VRML scenes by way of the Web via URLS.

VRML is also used as a file format for 3D objects and virtual worlds. It should not be confused with a modeling language. 3D objects and worlds are described in VRML files using a **scene graph**. This scene graph is composed of text. Text is combined to represent nodes which are composed to describe a specific item within a scene. Nodes can contain other nodes which create a scene graph structure and make it easier to create complex systems and subparts.

VRML has many object types. The simplest are known as **primitive objects** because they can be described in simple terms. These primitive objects include box, cone, cylinder and sphere which can be used together to make more complex objects. More complex shapes as well as surface materials and texturing can also be defined. Although VRML is not a typical programming language it still allows some of the similar benefits. Objects with a VRML scene can be linked to other URLs via hotspots. Other benefits include transformation. Transformation allows objects to be easily used more than once. A viewpoint setting allows users to look at values created within the scene or world. VRML creates a clear definition of lighting within the world and defines how particular objects will be rendered. VRML also works with other standard file formats in use on the Internet. Sounds, textures and animations can be linked to objects described in a VRML file by referring to image files, sound clips and program scripts in standard formats.

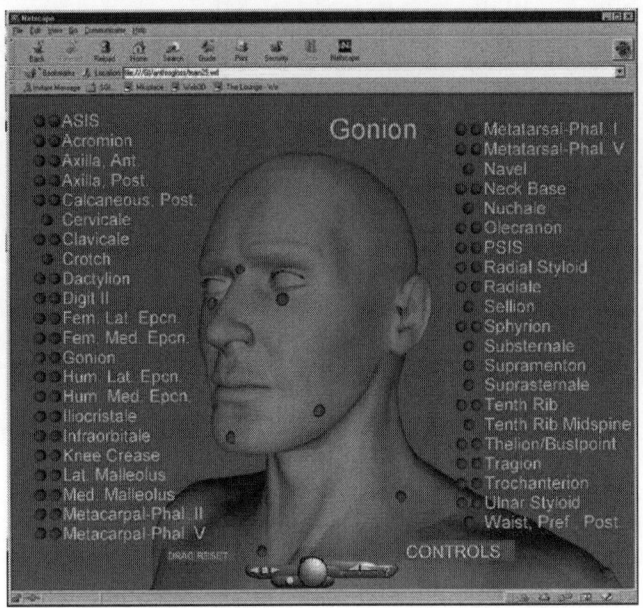

Example of a VRML primitive object

How to view VRML worlds

Most Internet browsers now include a VRML viewer which allows you to navigate and interact with a VRML world. A number of VRML viewers can be downloaded for little to no cost to the user, if needed. Viewers differ in the style of navigation and performance. VRML viewers may be selected based on familiarity with a particular style or form of navigation and are not dependent upon specific recommendation by a VRML developer.

Problems with VRML

VRML has distinguished itself as a pioneer to the 3-D world within the Internet. However, VRML is not without its' problems. VRML provides a variety of basic functions which are designed to run on all platforms. This has limited VRML capabilities from becoming the best all around virtual reality system. Other systems have been created for specific tasks or configured to run on specialized hardware. Limitations on VRML lighting and color have made it hard to overcome some of the more sophisticated, specialized virtual reality systems that continue to be developed.

Summary

VRML stands for Virtual Reality Modeling Language. VRML allows Web developers to create three-dimensional (3-D) space and 3-D objects in full color. VRML is also used as a file format for 3D objects and virtual worlds. VRML scene can be linked to other URLs via hotspots.

Concept Reinforcement

1. What does VRML stand for?

2. List four examples of primitive objects.

3. Explain two potential problems with VRML.

Section 1.15 – Organization of a VRML Scene

Section Objective

- Discuss the organization of a VRML Scene

Virtual Reality Modeling Language (VRML) was developed by the Web3D Consortium and was designed for use on the Internet. VRML is both a scene description language and a file format for virtual worlds. The language is used to describe the geometry and behavior of three-dimensional scenes. VRML is a popular 3-D format for the web because of its relative ease of use, standardization and relatively small file sizes that it produces.

What is a VRML Scene?

Programming languages such as C++, JavaScript, or HTML should not be confused with VRML. VRML is not a general purpose programming language, rather a scene description language. A scene can be described much like one from your favorite movie. A number of scenes are created to complete an entire movie. Within VRML, a number of scenes are combined to create a world. VRML is used to describe individual scenes by using a language that details the geometry and behavior within a 3D scene or world. More specifically, worlds received their name from an original goal of VRML which was to share worlds on the Internet. These worlds are made up of single files or groups of files. These files load individually at the same time to create a range of very simple objects to a complex scene.

VRML Scene Graph

A VRML text file describes a 3D world and the objects inside it. If a file is opened with a WRL (world) extension you will see a large amount of detailed text. That text is called a scene graph. A **scene graph** is the structure of the world being created. The VRML file format allows you to create a scene graph using words and punctuation. This is why you can create VRML scene graphs using a simple text editor like HTML.

Example of a 3D VRML scene created by Viktor Simov

A 3D world is defined one piece at a time in logical nodes. **Nodes**, which are part of a scene graph, describe a variety of items. Nodes may include a major feature, like a physical object in the world or even color of a specific object. VRML has three types of nodes. The first type of node is called a grouping node. Grouping nodes are used to contain and organize lists of other nodes. Children nodes are contained within these lists. Finally, attribute nodes are used within a specific field of a specific node. Attribute nodes specify detail such as shape or sound. As these nodes are grouped in their hierarchy, the scene begins to take shape. Each node changes or adds a specific attribute to the scene.

VRML is made up of more than 60 node types. A node is a piece that describes and controls some functionality. Within each node is a list of fields and events which hold values that define the parameters for that nodes function. Individual nodes are capable of fitting together in a logical order. The node above is called the **parent node** and the node below is called the **children node**. The relationship between parent nodes and children nodes is called the scene graph hierarchy. Nodes include a variety of potential values consisting of appearance, text, shape, sound, viewpoint, switch, as well as many others.

Summary

VRML is both a scene description language and a file format for virtual worlds. The language is used to describe the geometry and behavior of three-dimensional scenes. A scene graph is the structure of the world being created. The VRML file format allows you to create a scene graph using words and punctuation. A 3D world is defined one piece at a time in logical nodes.

Concept Reinforcement

1. Explain a scene graph as it relates to VRML.

2. What is the importance of a node?

3. What is the difference between a parent node and a children node?

Unit Two

Section 2.1 – Information Compression and Information Theory

Section Objective

- Explain Information Compression and Information Theory

Information Compression

Information compression is useful because it helps reduce the use of expensive resources and increases the speed at which you send and receive information. It can cut down on your hard disk and transmission space by minimizing the information being sent or saved. As with human communication, compressed information only works when both the sender and receiver of the information understand the encoding scheme. **Encoding** is the activity of converting data or information into code by using fewer bits while **decoding** is the opposite as it converts code back into the original information. For example, this text only makes sense if you understand the English language. Much like this text, compressed information can only be understood if the decoding method is known by the receiver. It is important to understand that information compression and information coding are two terms that generally mean the same. When information is compressed it is coded.

Compression is possible because information usually contains redundancies. Redundancies are information that is often repeated. Compression takes redundancies into consideration when encoding. For example, a straight black line contains redundant information in creating the color and shape from start to finish. Information compression will remove these redundancies by efficiently accounting for the entire line and color by recognizing that the patterns between shape and color do not change, therefore can be compressed.

As discussed, compression centers around two main applications, transmission and storage. Compression rate and compression ratio are two similar terms used to describe both. Compression rate calculates the speed at which information is to be sent from one computer to another. Typically this speed is measured in bits per second, bits per pixel, or simply, bits per character. Compression ratio calculates the size of the original data with the rate at which the data will be compressed. For example, if a simple gray image is represented by 8 bits per pixel and is compressed to 2 bits per pixel, we say the compression ratio is 4-to-1. This compression ration can also be represented as 75%.

Information Theory

The theoretical background or the study of information compression is known as the information theory. Information theory takes into consideration compression, communication, and error correction. Claude Shannon has been credited with creating the first fundamental papers on information theory in the late 1940s and early 1950s.

Shannon's definition of information revolves around identifying redundancy. If an information pattern repeats itself more often than other possible patterns of the same size then the information contains redundancy. If all the possible patterns of information occur equally then the information contains no redundancy. Shannon's information theory finds ways to compress redundant information. Eliminating information that repeats or is redundant is how we are able to increase the speed and response time of computer systems and cut down on storage space needed.

Lossless

Information compression incorporates different factors leading to the quality of decoding the original message exactly as intended. There are two main types of compression used, lossless and lossy. While done in two different manners, determining how information should be organized or coded and identifying redundancies is the goal of both lossless and lossy compression.

Lossless compression uses mathematical algorithms to calculate redundancy. An **algorithm** is a rule, or set of rules, used to calculate potential redundancy. This is used to gain the most accurate depiction of the original data being sent without error. We can use the alphabet as a simple example of lossless compression. In the English text, the letter 'a' is much more common than the letter 'z'. We also understand that the probability of the letter 'v' being followed by the letter 'q' is very small. Lossless compression takes this same type of general understanding and uses it to compress information. Lossless compression is a technique used to eliminate redundant information. Since only redundant information is removed it is often possible through lossless compression to restore the original information.

Lossy

Lossy compression is used when some loss of fidelity is acceptable in reducing the amount of data as much as possible. With lossy compression, redundant information is removed together with some of the non-redundant information. The loss of non-redundant information means that the original can never be perfectly restored. Lossy compression accepts this loss of data in order to achieve higher compression. We see lossy compression often with colors. The human eye is less sensitive to variations in colors, which allows compression to work by rounding off some of the less-important information.

Concept Extension: Data Intensive Computing

The technologies available to researchers create huge data sets, often resulting in terabytes of data produced by a single experiment. One of the challenges of these huge data sets is figuring out how to efficiently share and analyze the data scientists generate. The field of data intensive computing has developed to develop ways to capture, manage, analyze and understand very large volumes of data at high speed.

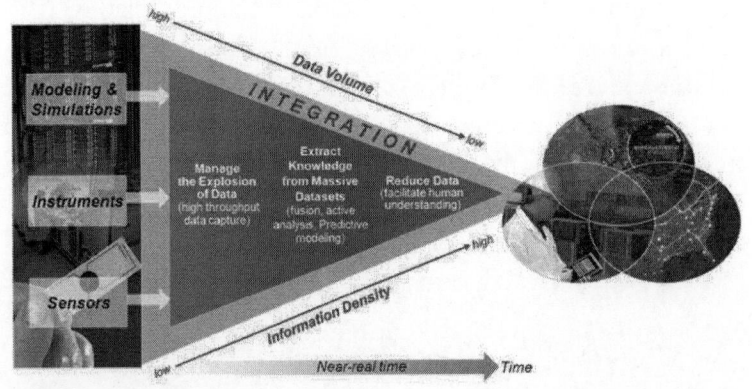

The Pacific Northwest National Laboratory is federally-funded laboratory that works in data intensive computing. PNNL is located in Richmond, Washington, and is funded by the Department of Energy. PNNL is one of the DOE's ten national laboratories, along with Argonne National Labs and others. The lab provides the expertise and resources required to strengthen the US ability to engage in fundamental research and innovation, conducts applied research in information analysis to increase security and counter terrorism and the proliferation of weapons of mass destruction, works to increase our understanding of hydrogen and biomass-based fuels to reduce our dependence on oil and reduce the effects of energy generation and use on the environment.

PNNL explores ways to manage the challenges presented by the increasing size of data files by conducting research in three primary areas:

- Managing the increasing amount of data

- Extracting knowledge from massive data sets. This is known as data mining.

- Reducing the amount of data to a point where people can understand and respond to it.

PNNL's approach to DIC: combine R&D in hybrid hardware architectures, adaptable software architectures, and advanced analytic algorithms to provide end-users with real capabilities that make impossible problems solvable.

Image courtesy of the Pacific Northwest National Laboratory, US Department of Energy

Summary

Information compression is useful because it helps reduce expensive resources and increases the speed at which you send and receive information. When information is compressed it is coded. The theoretical background or the study of information compression is known as the information theory. Information theory takes into consideration compression, communication, and error correction. There are two main types of compression used, lossless and lossy. Determining how information should be organized or coded and identifying redundancies is the goal of both lossless and lossy compression. Lossless compression is a technique used to eliminate redundant information. Lossy compression is used when some loss of fidelity is acceptable in reducing the amount of data as much as possible.

Concept Reinforcement

1. Briefly discuss information compression.

2. What role did Shannon play in the development of Information Theory?

3. What is the difference between lossy and lossless?

Section 2.2 – Probability-Based Coding

Section Objective

- Discuss probability-based coding and its uses

Code

Different codes and coding techniques are chosen based on the application and the desired amount of compression. Claude Shannon has been credited as the founder of information theory. He found that information can be related to probability. If an event is certain to occur then it has a probability of 100%. If the same event is not going to occur, it has a probability of 0%. We can attribute these percentages to binary to better understand the concept of probability coding.

If 100% represents the binary 1 and 0% represents the binary 0 then the probability of this event occurring takes place anywhere between 1 and 0. Let's say the event is a coin toss. The coin has a 50% chance of landing on either heads or tails. If a coin is tossed two separate times the probability of tossing two heads is: ½ x ½ = ¼ or 25%. In three tosses, the probability of tossing all tails is: ½ x ½ x ½ = 1/8 or 12.5%. Probability of the coin tosses is calculated this way because each toss is independent of the results of the other tosses. Probability coding works much like this example. The concept of probability coding is based on being able to convey information in fewer bits, on average, than it might first appear. If we know that the probability of a heads being tossed and a tail being tossed is 50% then we can eliminate all other possibilities from the equation.

Probability Coding

The mathematics of probability theory along with the science of statistics is the foundation of probability coding. Probability coding works to simplify data compression. It uses simplicity as a degree of data compression. The simpler the code, the more likely it will be for the code to be transmitted without any lost data. Probability coding is based on the ability to compress or simplify data by using probabilities. Probability coding works to play the odds. As a lossless compression technique, one file shrinks as another expands. It works to identify the probability of individual values within a random variable, or code. It also seeks the probability of the location within the code that the given value may fall.

Probability coding is based on familiarity. Files that are seen often are compressed while files that are seen rarely, if ever, are expanded upon. However, probability coding is only effective if there is a good idea of what kind of files are most likely to be transmitted. Those files that are more probable than others are exploited. For example, it is more probable that a string of letters used in the English language will show up as 'the' as opposed to 'xzq'. Probability compression thrives on this type of knowledge.

As humans, it is difficult to store every bit of detailed information we receive. Even if we could it would probably be of little to no use to us. Duplicated information stored in our brains would take too long to access in time to give a simple response to a question or thought. Because of this, we create short cuts. We choose what is important to us.

For example, we understand that most birds fly. We understand that the probability of a bird being able to fly is pretty high. We hypothesize these types of statements based on previous experience as key points. There is no reason to go through the day seeking to store the image or thought of every bird that we meet. We can think of our day as a huge set of data points, much as if our day has been coded. We work to compress all data that has been duplicated or repeated based on current or past information storage. Therefore, as we see birds flying throughout the day we 'compress' that specific information into a general understanding that most birds fly, which we already knew. Therefore, this duplicate information does not need to take additional storage space in our brain. However, a unique bird because of color or size may stand out to you. This bird may represent a data point that is new and would be compressed, interpreted, and stored as new data.

Summary

Different codes and coding techniques are chosen based on the application and the desired amount of compression. Shannon found that information can be related to probability. The mathematics of probability theory along with the science of statistics is the foundation of probability coding. Probability coding is based on the ability to compress or simplify data by using probabilities.

Concept Reinforcement

1. Briefly discuss probability coding.

2. What is a limitation of using probabilities in coding?

3. Explain how lossless compression affects file size.

Section 2.3 – Variable Length Coding

Section Objective

- Explain variable length coding and its uses

Coded information is based on the most efficient way of getting from point A to point B. As data is converted to binary we know that some of the information may occur more than once. Colors, fonts, patterns, or styles are all considerations when looking for repetitive data. It is possible to represent frequently occurring characters with a smaller number of bits during transmission. Morse Code is a good example of **variable length coding** in that it uses more efficient, shorter coding for symbols that are used most often. For example, an 'E' occurs more frequently than a 'Z' so 'E' is represented with a shorter code:

. = E

—.. = Z

This process of coding is represented by variable length code. Variable length coding maps symbols to a variable number of bits. Variable length coding is a form of lossless data compression because it is able to encode and decode with zero errors. Much like probability based coding, variable length coding systems will always assign the longest codes to the most infrequent events and give the shortest codes to the most frequent events. Variable length coding is best described by three well-known strategies: Huffman Coding, Arithmetic Coding, and Lempel-Ziv Coding.

Huffman Coding

Huffman coding is a statistical technique which attempts to reduce the number of bits required to represent a string of symbols. Huffman coding was developed by a Ph.D. student named David A. Huffman while at MIT. Symbols in Huffman coding vary in length to shorten the overall coding, which allows use of a more efficient compression technique. The idea behind Huffman coding is to simply use shorter bit patterns for more common characters, and longer bit patterns for less common characters. This often requires the assignment of codewords or patterns to the more familiar characters. This concept is similar to Morse code.

A seaman in the U.S. Navy sends Morse code signals.

Arithmetic Coding

Arithmetic coding is another useful technique for information compression. Arithmetic coding works a bit differently than Huffman Coding. Huffman coding is rather fast, but does not produce an efficient compression ratio. Arithmetic coding uses a coding table that is updated frequently to reflect real time distribution statistics. As a new character is being processed, the table will re-calculate until the end of the variable. This allows arithmetic coding to have a higher compression ratio than Huffman coding. However, these complex calculations take much more time than Huffman coding, resulting in slower transmissions.

Lempel-Ziv Coding

Lempel-Ziv coding is one of the most popular compression techniques for lossless data storage. Lempel-Ziv coding is a result of reusing frequently occurring variables. A compressed data string is always binary, meaning characters are coded using a series of zeroes and ones. It uses these binary digits to represent a previous string, then adds one new bit to represent a new data variable. This technique works best for long data streams and is actually very inefficient with short data streams. Lempel-Ziv is used by WinZip to compress larger amounts of information.

Summary

Information is coded is based on the most efficient way of getting the data from point A to point B. As data is converted to binary code, we know that some of the information may occur more than once. Variable length coding systems map symbols to a variable number of bits. Variable length coding is a lossless data compression technique because it is able to encode and decode with zero errors. Huffman coding is a statistical technique that attempts to reduce the number of bits required to represent a string of symbols. Arithmetic coding uses a table for coding that is updated frequently to reflect real time distribution statistics. Lempel-Ziv coding is a result of reusing frequently occurring variables.

Concept Reinforcement

1. Explain variable length coding.

2. Who is Huffman coding named after?

3. Explain how arithmetic coding differs from Huffman coding.

4. What does Lempel-Ziv coding work best for?

Section 2.4 – Universal Coding

Section Objective

- Explain universal coding and its uses

As discussed in previous chapters, file compression is a technology that reduces the size of data. Increased speeds of transmission and additional storage capacity are the two biggest benefits of compression. Compression typically allows the original data to be decreased in size by 1/5 to 1/2 that of the original. The goal of all compression types is to preserve all information and completely restore the original data without even a 1-bit loss or defect. **Universal coding** techniques have the capability to globally compress all types of data without loss. The disadvantage, however, is that universal coding is typically more complex than other compression technologies and has a slower overall speed in which data is transferred.

Universal code is used to efficiently communicate a message drawn from a set of possible messages. Universal coding works well when the probabilities of particular variables are unknown. It spots sequences that have previously appeared and uses this repeated information to build a more efficient message. Universal coding can be used with any combination of variables that repeat themselves in some form. The overall rule is to break up any combination that has been seen before into smaller stages. For example, a string of the letters X and Z in a message may look like this:

XZXXZXZXXZXXZ

Universal coding would look at the first string of letters that is new. In this case, the first variable is an X, which is new. This variable is then separated into its own stage from the rest of the group. That representation may look like this:

X/ZXXZXZXXZXXZ

The next variable is a Z, which has also not been seen before and would be separated into another individual stage:

X/Z/XXZXZXXZXXZ

Continuing, the next symbol in our string of XY variables is another X, which has already been seen. So, our next shortest variable string would be represented as XX. We can continue down the message and separate the new groups of variables into stages to look like this:

X/Z/XX/ZX/ZXX/ZXXZ

The universal compression method gains efficiencies by now looking at the entire message and understanding that each string may be very similar to one we've already seen. So, instead of describing the final sting as a new variable set it may recognize this string as 'ZXX' plus 'B' since both those individual strings have already been identified previously. So, rather than a new string of ZXXZ we may represent this as a combination of two strings that have been previously identified:

String 5 + Z

In this short example, we may not previously know the probabilities of X or Z occurring. Universal coding allows long messages to be efficiently transmitted. Universal coding does not require the coder to know the statistics of the frequency of the events to be coded. It is based on the concept that any stream of data consists of some repetition.

Why Universal Coding?

Different coding schemes can be used and offer different benefits depending on your overall expectations of the decompressed data. So, if we have codes that seem to accomplish the end result as desired, what is the purpose of universal coding?

Probability based coding and variable length coding can give as good, if not better, compression as any universal code. However, these codes cannot always be used, which is why the universal code is beneficial. For example, universal code can be used when the exact probabilities of a message are unknown. Universal codes can also be used when the transmitter and receiver are not able to communicate effectively. The transmitter may know the probabilities of the message, but the receiver is not able to decode the message. Universal code can also be used when the behavior of a source is not known in advance, or the behavior changes. Universal code does not have the constraints that the other codes may have. However, while a benefit in that respect; universal coding is not always the most efficient coding option.

Summary

Universal code can be used to efficiently communicate a message drawn from a set of possible messages. Universal data compression techniques have the capability to globally compress all types of data without loss. Probability based coding and variable length coding can give as good, if not better, compression that any universal code. However, these codes are not always able to be used which is why universal code is a benefit.

Concept Reinforcement

1. Explain universal coding.

2. What may be a disadvantage of universal coding?

3. Why would universal coding be used instead of other compression types?

Section 2.5 – Image Compression

Section Objective

- Explain image compression

Compressing an image is significantly different than compressing information. The concept of compression, of course, remains the same. **Image compression** is a technique used to make the file size of an image smaller. Depending on the compression used, the file size may decrease slightly or tremendously. Think of compression much like you would a balloon. Imagine that your balloon is filled with air. This balloon represents an image. If you squeeze all the air out of a balloon you are compressing the air to allow your balloon to shrink. It now becomes much smaller than the size of the original balloon. This is the same balloon that you started with but with much less air which would allow the balloon to squeeze into a number of small spaces that it otherwise would not have been able to do. It is still possible to reform the shape of the original air-filled balloon by simply blowing air back into it until you reach its original size.

A direct analogy to the balloon can be drawn with image compression. You start out with a very large file size of an image (the original air-filled balloon). By applying compression to the file (releasing air), the file shrinks to a fraction of its original size. You can now fit more images onto a storage device and transfer files at a much higher rate because the images are now compressed and take up less space.

The image on the left shows a picture of a flower that has been compromised in the compression process while the image on the right has been compressed perfectly to resemble the original image of the flower.

Lossless in Image Compression

Lossless compression returns to its exact original state. As with the balloon analogy, even though the air is out of the balloon, it is capable of returning to its original size and form. Images will look exactly the same before and after being compressed with a lossless compression scheme. The GIF format is the most common lossless compression used on the Web for images. Although it is lossless, it has the capability of showing a maximum of only 256 colors at a time. Cartoons are a good example of an image that is suited for the GIF format. Cartoons include many distinct lines and colors within an image which makes them much more likely to fit any GIF restraints. They may also be able to be saved with as little as 8 to 16 colors, which greatly decrease the required file size compared to a similar image saved with the 256 GIF color limit.

Lossy in Image Compression

Bits of information are permanently lost during compression and decompression of an image through **lossy** compression. Therefore, unlike the balloon analogy, an image will permanently lose some of the information that it originally contained. Fortunately, the loss is usually only minimal and hardly detectable. The most common image format on the Web that uses a lossy compression scheme is the JPEG format.

JPEG is a very efficient compressed image format. JPEG has the ability to show more colors within an image than GIF. JPEG is best used for images with many color variations, shading, or gradients. Photographic images are best used with lossy compression. JPEG is able to compress an image much more so than GIF because it is lossless and uses the flexibility of losing some minor bits of detail.

Both GIF and JPEG have their distinct advantages, depending on the types of images you are working with. It is often beneficial to save an image as both a GIF image and JPEG image. This allows you to determine which picture gives you the best quality at the lowest cost to storage space.

Summary

Compressing an image is significantly different than compressing information. Image compression is a technique used to make the file size of an image smaller. Images have certain properties that can be taken advantage of when encoding or decoding an image. There is even an opportunity to sacrifice some of the finer detail of an image depending on the final intentions. This saves bandwidth and storage space which in turn delivers your image quicker and takes up less storage space. There are two different types of compression, lossless and lossy. Lossless compression returns to its exact original state. Bits of information are permanently lost during compression and decompression of an image through lossy compression.

Concept Reinforcement

1. What is the difference between information compression and image compression?

2. Why would it ever be acceptable to sacrifice detail on an image?

3. What are the most common types of lossless image compression formats used on the Web?

Section 2.6 – Video Compression

Section Objective

- Explain video compression

Compressed information must be decompressed to then be used. This extra processing step is known to be detrimental to some information applications as data can be lost. Information can often afford for this to happen without a significant negative impact. Video compression works in a similar fashion in that video is also decompressed prior to being viewed. However, the video compression process works differently in one major way in that information cannot afford to be compromised during the compression process. It is not acceptable to a viewer to be watching a movie and have a scene deleted because the compression process wouldn't allow for it to remain. That being said, much is still being done to improve the overall speed of video compression.

Video Compression

Moving pictures date back to the 1890s. Technology has moved from filmstrip to magnetic tape to digital media capable of being transmitted anywhere in the world, including small hand held devises and mobile phones. Digital video started with video equipment developed by Bosch, RCA, and AMPEX. They each developed a version of a digital video recorder (DVR) in their research and development labs in the late 1970s. However, it was Sony that first came up with a commercial DVR in 1986. They created the Sony D-1. This was a format for uncompressed digital video recordings. This was a very expensive technology. It was generally only accessible to the professional television industry. Aside from the cost, the bigger problem with the Sony D-1 was that it worked only with uncompressed digital video. This created exceptionally large data files. Even with high-bandwidth limitations that we know of today with the internet, digital video can go nowhere without compression technology to make it storable and transferable.

Early movie projector using filmstrip to view video prior to digital technology

The foundation of video compression is Differential Pulse Code Modulation (DPCM). DPCM was developed in 1980. It breaks down video information into small packages of data. An analog signal is sampled. Sampling is the process of converting analog signals into a digital format. The video packages are analyzed at various points in time. The difference between the actual sample value and its predicted value is quantified, or specifically defined, and then encoded. This process allows a digital value to be formed. Mathematical equations are used to evaluate the difference between video frames.

Video codecs then evaluate the amount of data that needs to be digitized as well as the amount that needs to be predicted mathematically. Codec is the process of evaluating code as it is being decoded. The word comes from 'code-decode' and is used to evaluate accuracy of the video frames. The overall accuracy in which the digital samples are created to reflect their original analog form is based on the sampling rate and sample size. Sampling rate is the rate at which analog signals are converted into digital form while sampling size is the number of times a signal is sampled in order to accomplish the proper conversion.

Digital Video Formats

All commonly used digital video formats use some level of compression. Top of the line camcorders used for personal use to record home videos often use a digital 4:2:2 format. This format has minimal compression and a much higher digital sampling rate used to avoid compromising any lost video. Other camcorders use a 4:1:1 format, which depends on a greater level of compression, resulting in some minimal compromises in video quality.

A comparison between the 4:2:2 and 4:1:1 formats often reveal little or no difference to the untrained eye. However, when editing, copying, or working complex special effects, the 4:2:2 will reveal a much greater quality advantage. These benefits are often realized in a professional environment. When combining a widely-used compression method with the 4:2:2 format, you create one of the most popular professional formats. It is used more often for its practicality. It produces a high quality but still allows for technical ease of use.

Summary

Moving pictures date back to the 1890s. The foundation of video compression is Differential Pulse Code Modulation (DPCM). It breaks down video information into small packages of data. Even with high-bandwidth limitations that we know of today with the internet, digital video can go nowhere without compression technology to make it storable and transferable. All commonly used digital video formats use some level of compression. A comparison between the 4:2:2 and 4:1:1 formats often reveal little or no difference to the untrained eye. However, when editing, copying, or working complex special effects, the 4:2:2 will reveal a much greater quality advantage.

Concept Reinforcement

1. Explain Differential Pulse Code Modulation?

2. Describe the differences between the two main types of digital video compression formats, 4:2:2 and 4:1:1.

3. Explain the difference between sampling size and sampling rate.

Section 2.7 – MPEG Video Compression

Section Objective

- Explain MPEG video compression

MPEG stands for the Moving Picture Experts Group. MPEG was established in 1988 to develop standards for digital audio and video formats. The major advantage of MPEG compared to other video and audio coding formats is that MPEG files are much smaller yet still allow similar overall quality. This is because MPEG uses very sophisticated compression techniques.

Portable digital video device using MPEG video compression

MPEG video compression is used in many current and emerging products. Digital television sets, HDTV decoders, DVD players, video conferencing, Internet video, and other applications all incorporate MPEG video compression. These applications each benefit from video compression for a number of reasons. They may require less storage space for archived video information, less bandwidth for the transmission of video information, or a combination of both. For example, one format defined for HDTV broadcasting within the United States is 1920 pixels horizontally by 1080 lines vertically. When you consider the amount of frames per second as well as the bits used for each primary color, the total data rate required becomes extremely high. This signal would need to be compressed by a large figure. MPEG allows these large compression rates to occur with as little visible loss as possible. This becomes even more impressive when you consider the high quality video offered to viewers in many homes today.

The International Telecommunications Union (ITU) and the International Standards Organization (ISO) worked together to standardize digital video compression. In 1993 the MPEG-1 format was developed. This allowed digital video and related audio to be stored on a CD-ROM. In 1994, the MPEG-2 format was created. This opened up the doors to make high definition television (HDTV) and DVDs possible. MPEG-4 was introduced in 1999 and allowed multimedia to be viewed over the Internet.

MPEG Video Layers

MPEG video is broken up into layers to help with error handling, random search and editing, and synchronization. The first layer, or the top layer, is known as the video sequence layer. This is any self-contained bit stream such as a movie or advertisement. The second layer down is the group of pictures. The third layer down is the picture layer itself. This is followed by the slice layer.

MPEG Compression Standards

MPEG standards are each designed with a specific application and bit rate in mind. There are a number of MPEG standards that are either currently used or in development today. The MPEG-1 compression standard is primarily used for moving pictures and audio. This was based on CD-ROM video applications. It has become popular on the Internet. The popular MP3 standard for digital compression for audio files comes from level 3 of MPEG-1.

Digital television and DVD compression is based on the MPEG-2 compression standard. It was developed from MPEG-1, but designed for the compression and transmission of digital broadcast television. It is able to more efficiently compress video over MPEG-1 and works will with HDTV resolution and bit rates.

MPEG-4 has become the standard for multimedia and Web compression. It is object-based, similar to VRML. A MPEG-4 file is created by taking separate objects within a scene and compressing them. This allows developers to control objects independently within a scene, which allows them to allow objects to interact. There are a number of other standards that have been released, are currently under development or have been slated to be released in the future.

Summary

MPEG stands for the Moving Picture Experts Group. The major advantage of MPEG compared to other video and audio coding formats is that MPEG files are much smaller yet still allow similar overall quality. MPEG video compression is used in many current and emerging products. The International Telecommunications Union (ITU) and the International Standards Organization (ISO) worked together to standardize digital video compression. MPEG video is broken up into layers to help with error handling, random search and editing, and synchronization. MPEG standards are each designed with a specific application and bit rate in mind.

Concept Reinforcement

1. What is the advantage of MPEG?

2. Which two groups worked together to *standardize digital video compression?*

3. Describe the differences between MPEG-1, MPEG-2, and MPEG-4.

Section 2.8 – Digital Television

Section Objective

- Describe digital television

The advancements of digital television (DTV) are working to make a direct impact on every household across the country. It is an advanced over-the-air broadcasting technology that will improve the average television viewing experience. DTV is more efficient and flexible than the traditional broadcast technology known as analog. It enables broadcasters to offer television with crystal clear pictures and CD quality sound. Interactive video and data services are also possibilities that DTV may offer over analog technology.

The Transition to Digital Television

In 1996, the U.S. Congress authorized the distribution of an additional broadcast channel to each broadcast television station. This allowed individual stations to start a digital broadcast channel while continuing with an analog broadcast at the same time. Congress later set June 12, 2009 as the final date that full power television stations can broadcast analog signals. They will now all be required to broadcast digital, over-the-air signals.

The decision by Congress to allow stations to broadcast in digital signals will help free up parts of the valuable broadcast spectrum. This spectrum allows police, fire departments, and rescue squads the ability to communicate uninterrupted. Additional space made available by the transition to digital will be auctioned to companies that will be able to provide consumers with more advanced wireless services.

Digital television is much more efficient than analog. Television stations are able to broadcast multiple channels of programming at the same time instead of simply being able to offer one. For example, a station broadcasting channel 12 in analog is only able to offer the viewer one program. A station broadcasting in digital on channel 12 can offer viewers one digital program on channel 12-1, a second digital program on channel 12-2, a third digital program on channel 12-3, and so on. This provides more programming choices for viewers. The ability to offer multiple programming choices is called **multicasting**. Multicasting also provides the flexibility for the broadcaster to offer a high definition digital program or multiple standard definition digital programs.

DTV has the ability to offer a picture in high definition. However, DTV should not be confused with **HDTV**, high definition television. HDTV is the highest quality of DTV but is a separate format. It offers the best available picture resolution, clarity and color. It also provides theatre surround-sound and a wide screen format which gives a 'movie-like' format to the viewing experience. **SDTV**, standard definition television, is another common format. Having DTV does not automatically allow a consumer to view a television picture in high definition. Specific television sets are needed to receive high definition television programming.

Television such as this HDTV are quickly gaining popularity

Buying a Digital TV Set

There are three primary types of television sets that can be found in an electronics store today. Increasing options make it important to have a general understanding of what you expect out of your television prior to making a new purchase. Analog television sets are still available. Remember, stations will now only broadcast digital, over-the-air signals, making an analog television obsolete on its own. Digital-ready sets are also available to consumers. These televisions are capable of handling the digital transition and will be ready to view any DTV broadcast. However, these televisions are not capable of viewing a broadcast in high definition. Since technology is moving broadcast television quickly towards these higher quality pictures, a digital-ready set may not be the most efficient purchase for the future. HDTV-ready sets, however, do provide the option of watching a higher resolution broadcast in high definition.

Summary

The advancements of digital television (DTV) are working to make a direct impact on every household across the country. DTV is more efficient and flexible than the traditional broadcast technology known as analog. It enables broadcasters to offer television with crystal clear pictures and CD quality sound. In 1996, the U.S. Congress authorized the distribution of an additional broadcast channel to each broadcast television station. Congress later set June 12, 2009 as the final date that full power television stations can broadcast analog signals.

Concept Reinforcement

1. What is the difference between DTV and analog television?

2. Explain the difference between HDTV and SDTV.

3. What type of options are available when shopping for a television?

Section 2.9 – The Theory of Sound

Section Objective

- State the theory of sound and discuss its importance

The science of sound is known as acoustics. **Acoustics** is created by both natural causes as well as by human activity. It also includes the sensation of hearing. Speech, music, sound recording and reproduction, telephony, noise control, as well as a number of additional examples each are associated with sound and hearing, or acoustics.

Sound is a means of transmitting information. Sound has been researched over time through our natural ability to hear. Studies of sound include the occurrence throughout air that is responsible for the sensation of hearing. Disturbances with frequencies too low or too high to be heard by a normal person are regarded as sound. These low frequencies are known as **infrasound** while the high frequencies are known as **ultrasound**. It has also been closely studied to determine how sound travels and how sound is interpreted based on the medium through which it passes. The physical effects of sound are all studied through different theories.

One theory revolves around waves. A **sound wave** is a pattern of disturbance caused by movement. Movement is measured by an energy which occurs within a medium such as air, water, or any liquid or solid matter. Sound waves are heard as a source is disturbed or vibrates. This can be a ringing telephone or a voice of a person. These sounds disturb a particle in the surrounding medium and those particles disturb those next to them and so on as they travel through the medium. This disturbance creates movement in a wave pattern much like waves of water. The wave then carries the sound energy through the medium until it is received by the source.

The theory of sound dates back to at least 240 B.C. as the Greek philosopher Chrysippus, the Roman architect and engineer Vetruvius, and the Roman philosopher Boethius each theorized that sound movement might take a wave form. This is our first documented theory of sound. Their theory of sound in a wave form reflected around a number of rules. These rules revolved around sound and reflection. They believed that sound spread spherically however water waves spread out in a single plane. This caused conflict in theories and made individual rules difficult to explain.

Another theory of sound accepted today is that of Isaac Newton. Isaac Newton was born in 1643. He was a physicist, mathematician, astronomer, philosopher, and theologian. He is known as one of the greatest geniuses to ever live and is best known for developing a variety of accepted laws and theories, most notably that of gravity.

Issac Newton

Isaac Newton developed a mathematical theory of sound in 1687 as documented within his book, Principia. He believed that sound could be interpreted as pressure pulses that are transmitted through side-by-side particles. However, his theory was incomplete and left a number of approximations and incomplete definitions. It eventually became recognized as a legit theory of sound once studied and completed.

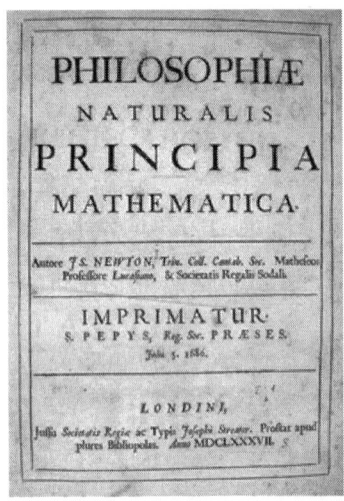

Title Page of Issac Newton's, *Principia*

The eighteenth century saw a productive leap towards a viable theory of sound. This theory included mathematical and physical concepts made by Euler Lagrange and d'Alembert. While incomplete from many standpoints, they were still able to combine a mathematical structure to sound that eventually led to many of the modern day theories of sound accepted today.

Summary

The science of sound is known as acoustics. Acoustics is created by both natural causes as well as by human activity. Sound is a means of transmitting information. The physical effects of sound are all studied through different theories. One theory revolves around waves. A sound wave is a pattern of disturbance caused by movement. The theory of sound dates back to at least 240 B.C. as the Greek philosopher Chrysippus, the Roman architect and engineer Vetruvius, and the Roman philosopher Boethius each theorized that sound movement might take a wave form. Isaac Newton developed a mathematical theory of sound in 1687 as documented within his book, Principia. He believed that sound could be interpreted as pressure pulses that are transmitted through side-by-side particles.

Concept Reinforcement

1. Explain infrasound and ultrasound.

2. Briefly discuss the theory of sound that revolves around waves.

3. What is Isaac Newton's theory of sound?

Section 2.10 – How Sound is Converted to Signals

Section Objective

• Explain how sound is converted into a signal

Sound waves are created in different formats. These formats are known as audio signals. This is the basis for sound conversion into a signal. Techniques and tools are used to simplify this process. Signals can be expressed through a number of different mediums such as electronic voltage, magnetic particles, radio frequency waves, or even pulses of light. It is important to convert sound into an audio signal to be able to manipulate, store, transmit or even reproduce the original sound. Signals allow sound to be manipulated in a way that sound waves themselves are not able to accomplish. To measure electric energy in an audio signal, decibels are used in relation to either power or voltage.

Sound needs to be converted into a signal in order to save, or record. Recording sound is typically done by using a microphone. Microphones are able to convert sound pressure waves into voltage. This creates an electrical audio signal. In order to convert sound into an electrical signal a transducer is used. When working within a specified electrical range, transducers are often called microphones. Sound varies in output depending on what is being measured. A human speaking directly into a microphone will need to be measured differently than the sound of a hummingbird from several feet away. In order to record and manipulate all ranges of sound, an amplifier is used. An **amplifier** allows specific signal amplitude to be measured and manipulated.

All microphones and transducers have a range of sound that can be manipulated and recorded. This range is known as the **frequency range**. If sound falls outside the frequency range it becomes unusable. **Dynamic range** is another consideration when converting sound to a signal. If the sound is too loud or too soft it may not be able to reproduce the sound reliably once converted to a signal. Range of sound is known as **amplitude**. Understanding the type of sound you choose to work with will help in making a choice between the type of microphone or transducer you will need. Once sound has been converted to a signal it can be recorded. Recorded sound can then be played back through speakers or headphones to convert the electrical audio signal back into sound.

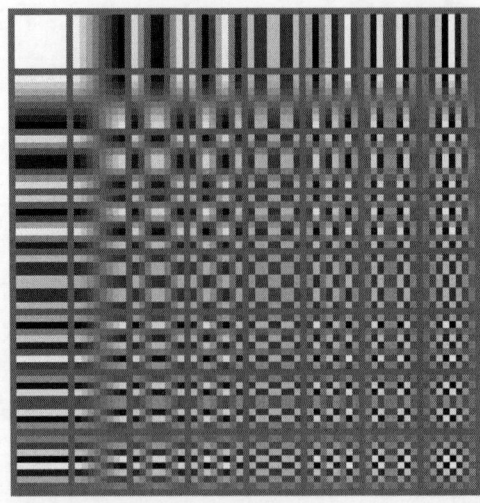

An image of the basis functions of the discrete cosine transformation used
by jpeg image file format and various video encoding standards

Once sound is recorded it can be converted to a digital signal and saved directly to a computer. This allows the user to manipulate, store, or send sound signals. In this case, the analog sound signal must be converted into a digital signal. One common technique used to accomplish this is called the **Pulse Code Modulation (PCM)**. PCM is based on a sampling rate needed to convert an analog signal to a digital signal. **Sampling** is the process of replacing portions of the analog signal with amplitudes taken at a regular interval. These amplitudes are measured and given a valid value. The sampling rate allows the original sound to be created again when necessary.

Summary

Sound waves are created in different formats. These formats are known as audio signals. This is the basis for sound conversion into a signal. Signals can be expressed through a number of different mediums such as electronic voltage, magnetic particles, radio frequency waves, or even pulses of light. It is important to convert sound into an audio signal to be able to manipulate, store, transmit or even reproduce the original sound. Recording sound is typically done by using a microphone. Microphones are able to convert sound pressure waves into voltage.

Concept Reinforcement

1. What different mediums can signals be expressed through?

2. Why is it important to identify the frequency range when using a microphone?

3. Explain how a signal is converted digitally through Pulse Code Modulation.

Section 2.11 – Frequency and Bandwidth of Audio Signals

Section Objective

- Discuss the frequency and bandwidth of audio signals

Audio Signals

A sound wave is an analog signal as it assumes continuous values throughout its length of time. Sound waves are represented in different forms through audio signals. Audio compression and coding techniques are used to compress audio signals. Audio signals can be represented as an electrical voltage or through a number of different mediums including radio, pulses of light or even through fiber optic cables. These signals can be stored, transmitted and reproduced. Audio signals can be measured by using a mathematical technique. However, in order to measure audio signals, frequency and bandwidth must be accounted for.

Frequency is a measurement of the number of cycles repeated per event in a given time. In measuring the frequency of audio signals, electromagnetic waves, or signals, are measured in number of cycles per second. The overall range of these radio or light frequencies, or the difference between the highest and lowest frequency measured, is known as **bandwidth**.

Frequency

The vibration of material causes sound. Sound is transmitted through a medium, such as air, as pressure waves are formed around the vibrating material. For example, strings of a guitar vibrate when stroked upon. Pressure waves create a pattern referred to as the **waveform**. These waves occur at regular intervals of time called a **period**. The amount of periods per second is known as the frequency of audio signals. These are measured in hertz (Hz) or cycles per second (cps).

Strings of a guitar vibrate when stroked.

Frequency can be calculated by counting the number of periods in a specified amount of time. This number is divided by the length of the time interval. The result is presented in Hz. One hertz is an event that occurs once per second. The human ear can typically recognize frequency in a range of 15 to 20,000 hertz. This is designated as VLF, very low frequency, and is defined as the audio frequency range. The human voice, on the other hand, can be characterized in the 50 to 10,000 hertz range.

Bandwidth

The importance of bandwidth is vital when transmitting audio signals. For an audio signal to be successfully transmitted, the medium bandwidth must be equal or greater than the actual signal bandwidth. Frequency of a signal may be negatively compromised, or lost, if the medium bandwidth is lower than the signal bandwidth. This certainly lessens the overall quality of the signal. Loss of quality is caused by a bandlimiting channel. Selecting a medium bandwidth large enough to fit the audio signal is vital towards successful audio signal transmission. This often requires a compression technique to allow the overall bandwidth to be manageable.

Summary

Audio compression and coding techniques are used to compress audio signals. Audio signals can be represented as an electrical voltage or through a number of different mediums including radio, pulses of light or even through fiber optic cables. In order to measure audio signals, frequency and bandwidth must be accounted for. Frequency is a measurement of the number of cycles repeated per event in a given time. Bandwidth is the difference between the highest and lowest frequency measured.

Concept Reinforcement

1. What unit of measurement is frequency measure in?

2. Why is would compression techniques ever be needed when transmitting audio signals?

3. How are frequency and bandwidth utilized when transmitting audio signals?

Section 2.12 – Sampling of Audio Signals

Section Objective

- Explain how to sample audio signals

Sampling and Sampler

Analog signals differ from digital signals in that analog signals are typically continuous in time. Digital signal processing does not use the entire analog signal. It replaces it by its amplitudes taken at regular intervals and works to reduce the continuous signal to an individual, isolated signal. This is **sampling**. The trick to sampling is in trying to reconstruct the original analog signal perfectly. Sound waves are often sampled into individual sequences when converting to a digital format. The process of sampling centers on a combination of individual samples. A **sample** refers to a value or set of values at a point in time. A **sampler** is used to extract samples from a continuous signal.

> In digital audio, common sampling rates are:
>
> **8000 Hz** - telephone
>
> **22050 Hz** - radio
>
> **44100 Hz** - compact disc
>
> **48000 Hz** - digital sound used for films and professional audio
>
> **96000** or **192400 Hz** - DVD Audio

Much of the world of sampling revolves around digital signals. This is referred to as digital signal processing. Through digital signal processing, the **sampling frequency** is measured. This is the rate at which an analog signal is sampled into a digital signal consisting of digital samples. Sampling frequency is usually measured in samples per second, or hertz. This measurement is often referred to as the **sample time**. Sample time is either the time a particular sample was taken or the period of samples taken from a specific signal. Typically, higher frequencies result in higher-quality sampling.

Nyquist Sampling Theorem

Sampling frequency depends on the signal being sampled. However, there are limitations on what can be measured. These limitations are based on the Nyquist Sampling Theorem. This theorem was developed by Harry Nyquist. Harry Nyquist was a physicist who worked at AT&T and Bell labs. In 1927 he developed a study centered on the sampling rate needed for an analog signal to be converted to a digital signal. The Nyquist Sampling Theorem states that the sampling frequency must be greater than twice the bandwidth of the input signal. This allows the signal to be reconstructed to its originally state successfully. The signal may be distorted if the sampling is performed at a frequency lower than the Nyquist Frequency.

For example, if the maximum frequency of an audio signal is 20kHz, this signal would need to be sampled with a sampling frequency of at least 40kHz. This allows the original signal to be captured in the sampled values so it can be perfectly recreated.

Analog-to-Digital Converter

Sampling is typically followed by two additional operations. These operations include quantization and binary encoding. Both work to minimize the overall size of the signal and allow the sound to be as efficient as possible without compromising the overall sound quality. All three steps are accomplished by an analog-to-digital converter (ADC). The job of the ADC is to store the value of the signal as a digital number. Once the signal is sampled by the ADC, a digital to analog converter (DAC) can be used to reverse the process. The DAC uses the digital signal to reconstruct the sound back into an analog form.

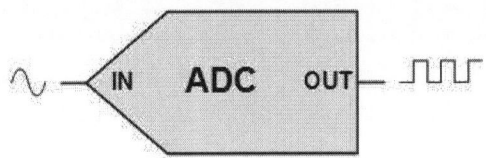

ELECTRICAL SYMBOL FOR ANALOG TO DIGITAL CONVERTER (ADC)

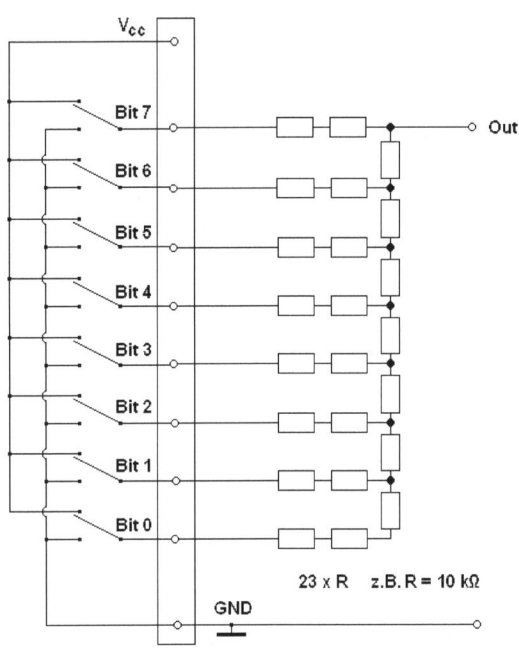

A scheme for a digital analog converter

Summary

Sound waves are often sampled into individual sequences when converting to a digital format. The process of sampling centers on a combination of individual samples. A sample refers to a value or set of values at a point in time. This is referred to as digital signal processing. Through digital signal processing, the sampling frequency is measured. This is the rate at which an analog signal is sampled into a digital signal consisting of digital samples. The Nyquist Sampling Theorem states that the sampling frequency must be greater than twice the bandwidth of the input signal.

Concept Reinforcement

1. Explain the importance of sampling an audio signal.

2. Explain *the Nyquist Sampling Theorem.*

3. What is the difference between a ADC and DAC?

Section 2.13 – Digitizing Audio Samples

Section Objective

- Describe the process of digitizing audio samples

The human ear is able to interpret audio samples by picking up vibrating pressure waves. These waves move in a pattern called waveform. To simplify, sound can be stored by waves which are analog signals. When audio samples are needed to be used in any computer application they must be converted into an electrical, or digital, signal comprised of a binary stream of 0s and 1s. The process of converting analog signals to digital signals is called **digitizing**.

Once audio signals have been digitized they can become a major element of multimedia. **Multimedia** is the use of computers to present text, graphics, video, animation or sound in an integrated way. Adding appropriate audio samples to a Web page can enhance the overall product. This can increase the opportunity for learning, comprehending or interacting with enhanced activities. This is a fairly simple process when a sound file or audio sample has already been digitized. However, when sound files are unavailable they must be created. Digitizing sound is a vital part of the overall process of creating an audio sample.

A female HDMI connector, which is used in digitizing files

The audio sample must first be recorded. It can be recorded to a number of options including a CD or digital media device. There are a number of parameters and considerations needed when digitizing sound that will impact the amount of information that is available to be stored in a file and the overall quality of the digital sound. Sampling rate, bits per sample, and mono vs. stereo must all be understood.

The **sampling rate** is the number of times the analog sound is sampled during each period and converted into digital information. The more samples taken, the closer the digital version will resemble the original analog version. **Bits per sample** describes how much information in each sample the computer is collecting. Bits per sample converts a sampled sound into an equivalent digital value. Sound quality and options increase as more bytes are used. It is important to balance the sampling rate and bits per sample to come up with an acceptable quality of the sound with the minimum file size.

Mono describes a sound system where all audio signals are mixed together and routed through a single audio channel. Stereo sound systems have two independent audio channels. The signals in a stereo sound system are reproduced by two channels. Two sound channels give the illusion that sound is coming from a certain location. While stereo systems are often seen as the most popular choice, mono systems are able to reduce the file size by half. Depending on the quality of systems, a mono system can sometimes be better than a low quality stereo system.

Equipment Use to Digitize Sound

In order to properly digitize audio samples you must have the proper equipment. An audio source such as a CD is needed to house the original digital source. A computer will also be needed to house the transferred sound from the audio source. Transferring sound from an audio source to a computer is done via an audio cable. Software will be helpful in allowing the user to digitize and edit sounds once stored onto the computer. Headphones or speakers are helpful towards allowing the sound to be adjusted or edited.

Summary

The process of converting analog signals to digital signals is called digitizing. Digitizing sound is a vital part of the overall process of creating an audio sample. There are a number of parameters and considerations needed when digitizing sound that will impact the amount of information that is available to be stored in a file and the overall quality of the digital sound. Once audio signals have been digitized they can become a major element of multimedia. Multimedia is the use of computers to present text, graphics, video, animation or sound in an integrated way.

Concept Reinforcement

1. What is the process of converting analog signals to digital signals called?

2. How does the human ear interpret sound?

3. Explain the equipment needed to digitize sound.

Section 2.14 – Quantization of Noise

Section Objective

• Describe the quantization of noise

MP3s or JPEG images are downloaded, saved, transferred or edited frequently by many people across the world. These formats would not be nearly as efficient or easy to work with if it were not for quantization. **Quantization** is the process of approximating a specific range of values within a signal. A small set of distinct symbols are used to identify these values. Quantization works along with compression to lessen the overall size of files.

Quantization and Data Compression

Lossless compression and lossy compression can often be distinguished by quantization. Quantization plays a major part in lossy compression. It is often motivated by the need to reduce the amount of data needed to represent a signal. As with the popular MP3 format, compression is achieved by removing some data which can be considered a lossy process. This process is often referred to as quantization.

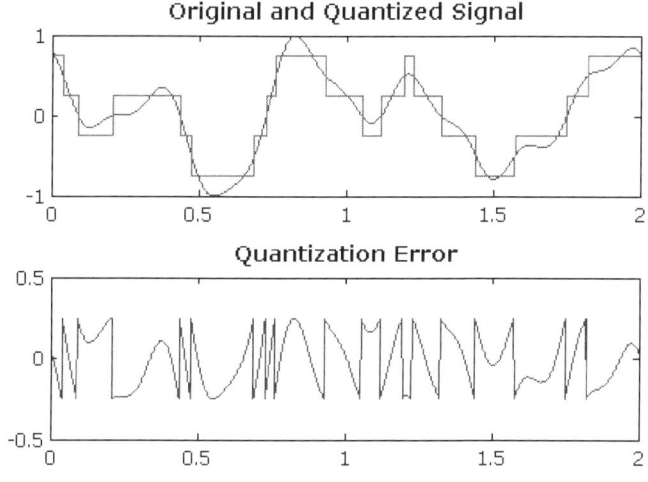

Plot of a quantized signal and its error.

A digital-to-analog conversion process occurs at sampling intervals. It always involves some amount of approximation, or rounding, as the amplitude of the waveform is quantized. The quantized process involves converting each sample amplitude value to the nearest approximate binary values. Once the approximated digital values are converted back into the original analog signal, the impact of the approximations equates to quantizing noise. If the sampling rate is too low or the approximation is too much, the reconstructed signal could become too inaccurate to consider as acceptable to the user due to the poor quality. The potential loss of data is how the lossy compression technique became associated with quantizing.

JPEG image compression is a popular example of a lossy compression scheme that uses quantization. When a JPEG image is encoded, the data representing the image is processed and then quantized and coded. The number of bits needed to represent the image can reduce the precision of the transformed values. For example, images can often be represented using the JPEG format in less than three bits per pixel as opposed to the typical 24 bits per pixel needed prior to JPEG compression.

Summary

Quantization is the process of approximating a specific range of values within a signal.

Quantization works along with compression to lessen the overall size of files. As with the popular MP3 format, compression is achieved by removing some data which can be considered a lossy process.

Concept Reinforcement

1. Describe the overall goal of quantization?

2. Compare quantization to lossy compression.

3. Explain how JPEG compression uses quantization.

Section 2.15 – CD Players

Section Objective

- Describe CD players

Music has always been a popular source of home entertainment. The invention of the phonograph in 1876 started the popularity of at-home music listening. In recent years, the compact disc player has become the medium of choice for much of the recorded music we listen to on a daily basis in our cars, homes, while exercising, or even at work. The CD and CD player have expanded their uses outside of simply the music industry into one of the most widely used devises and medium options for all digital data storage.

A **CD player** plays audio CDs. It is known as an optical medium. It uses a tightly focused light source called a laser to read or detect information. A **CD**, or compact disc, is an optical storage medium with digital data recorded on it. The form of digital data within a CD is not simply limited to music. CDs are used to record music, video or computer information. CD players have become flexible in its variety of uses. It can be found as an individual component within a stereo system. These often contain a drive and electronics needed to decode the digital sound. A CD player can also be a handheld, portable device. These often require headphones for use. CD players can also be self-contained units that include an amplifier and speakers.

History of CDs

The history of compact discs can be traced back to the 1960s when the development of digital electronic technology began to take shape. By the 1970s, digital and optical technologies started to combine to develop a single audio system. The successful creation of the CD helped conquer the three main challenges faced by developers of digital audio. The challenges included finding an efficient method for recording audio signals in a digital format. The second challenge centered on finding a suitable storage medium to accommodate this method. Finally, the medium needed to be able to process this large amount of information quickly enough to produce continuous music. By the end of the decade a common set of standards for optical storage discs had been developed by joint efforts of Sony and Philips. The standard was adopted in 1981 and the first CDs and CD players were introduced in the market in 1982.

How CDs Work

A CD is 4.72 inches in diameter, .047 inches thick, and weighs about .53 of an ounce. While seemingly simple devices, the CD actually requires a significant amount of technology to create each individual disc. Three physical layers are combined to make each CD. The base layer is made up of plastic. A thin layer of aluminum coating is added to the top of the plastic followed by a clear protective acrylic coating over the aluminum layer.

Information is recorded on the underside of a CD. Recorded information takes the form of a continuous spiral starting from the inside of the disc and moving towards the outer edge. These spirals are made up of pits and lands. Pits are a series of indentations within the spiral while land is the section that separates each pit. A tiny laser beam moves along the spiral track and reflects light back to a photo sensor. The variations in light represent the information originally recorded.

As with any successful and popular technology, the CD and CD player have continued to be a popular choice by consumers. The CD is a dependable storage device that is relatively immune from wear and tear if cared for properly. Since the CD is a very precise and accurate devise, it is not only important for the manufacturing process to eliminate any potential errors or flaws within a disc but for the user to properly care for the disc as well. The information within a disc is small enough to allow error with even the smallest of dust particles. These can even render a disc unreadable.

Summary

A CD player plays audio CDs. It is known as an optical medium. A CD, or compact disc, is an optical storage medium with digital data recorded on it. The form of digital data within a CD is not simply limited to music. CDs are used to record music, video or computer information. The compact disc standard was adopted in 1981 and the first CDs and CD players were introduced in the market in 1982. A CD is 4.72 inches in diameter, .047 inches thick, and weighs about .53 of an ounce. Three physical layers are combined to make each CD.

Concept Reinforcement

1. Explain the different types of CD players.

2. What are the dimensions of a CD?

3. When were CDs and CD players introduced to the market?

Unit Three

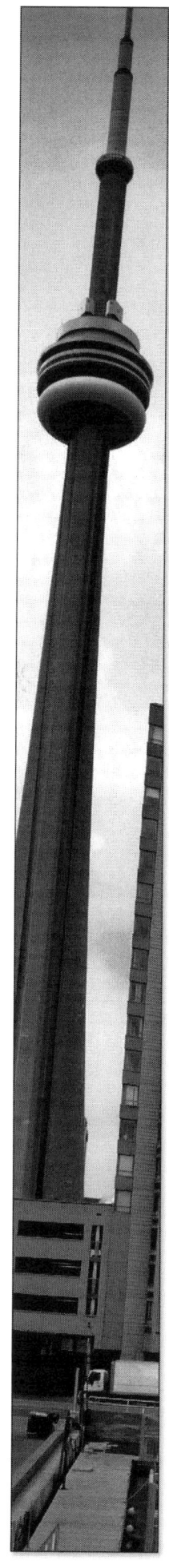

Section 3.1 – Analog Telephones

Section Objective

- Explain how analog telephones function

Analog

A telephone is a communication device capable of transmitting sounds over large distances. Telephones became a reality in the mid-19th century. However, for those born after the 1970s it may be difficult to remember a time when analog telephones were the standard rather than digital. Improvements and progress in technology and communication fields have led towards the shift from using electronic pulses (analog) to using bits of data (digital) to transmit sounds through a telephone. Prior to the change in primary telephone communication, analog telephones were the original telephone technology.

Early example of an analog telephone dating back to 1896

Scottish inventor, Alexander Graham Bell, received the first patent and was credited with the invention of the telephone at age 29 in 1876. In 1877 Alexander Graham Bell formed the Bell Telephone company. Bell was a driven, curious and intelligent man who did not let the success of the telephone slow him down. He later spent a good majority of his life dedicated to such concepts as kites, airplanes, architecture, animal breeding, artificial respiration and water distillation.

Alexander Graham Bell is credited with the invention of the telephone.

Analog Cell Phones

We often automatically assume all cellular phones are digital. Well, much like the telephones connected to your LAN (Local Area Network) line, cell phones were once primarily analog. The FCC approved the analog cell-phone standard called AMPS in 1983. AMPS stands for Advanced Mobile Phone System which was first used in Chicago. 824 megahertz (MHz) to 894 megahertz is the frequency range used for analog cell phones within AMPS.

The U.S. government was determined to encourage competition within the analog cell phone industry in order to keep prices low so they required two carriers to be available within each geographic market. These carriers were known as A and B carriers. One of the carriers typically ended up being the local-exchange carrier (LEC). The local telephone company is often referred to as the LEC.

Each carrier, A and B, were assigned 832 frequencies. 790 frequencies were regulated for voice while the remaining 42 were set aside for data. Two frequencies are combined to make a pair. One frequency is used for transit while the other is for receiving. Each pair is combined to create a **channel**.

NAMPS is a version of AMPS. NAMPS (Narrowband Advanced Mobile Phone Service) differs from AMPS in that it incorporates some digital technology to allow the system to carry about three times as many calls as the original AMPS version. Although it uses digital technology it is still considered analog. NAMPS operates in the same frequency band as AMPS and does not offer many of the features found in digital cellular service today, such as e-mail and Web browsing.

There is often a misconception between analog and digital telephones and how they differ. Many assume the difference to be the power source or electric current required for a digital telephone. However, the actual distinction between the two types of telephones pertains to the manner in which sound is transmitted. Analog telephones convert air vibrations into an electrical analog frequency. Analog technology manipulates sound waves allowing them to transform into electric pulses that are then able to travel between telephone devices. Digital telephones also transforms sound waves but in the form of binary data.

Summary

A telephone is a communication device capable of transmitting sounds over large distances. Scottish inventor, Alexander Graham Bell, received the first patent and was credited with the invention of the telephone at age 29 in 1876. The FCC approved the analog cell-phone standard called AMPS in 1983. Analog telephones convert air vibrations into an electrical analog frequency. Analog technology manipulates sound waves allowing them to transform into electric pulses that are then able to travel between telephone devices. Digital telephones also transforms sound waves but in the form of binary data.

Concept Reinforcement

1. Describe the difference between analog and digital as related to telephones.

2. Who is credited with the invention of the telephone?

3. What is the difference between AMPS and NAMPS?

Section 3.2 – Digital Cell Phones

Section Objective

- Describe how digital and cellular phones operate

Digital

Cellular technology, analog or digital, is often referred to as telephones. However, they are more accurately referenced as a highly sophisticated radio. Cell phone towers are also radio towers. Cell phone company towers located throughout the country are known as their individual cellular network. While the tower may be the same, the technology is much different. A cell phone can share the same bandwidth used by other cell phones because its signal is very weak. However, a radio station transmits a very strong signal in order to be heard over a large area, making it impossible to share the same frequency.

An example of a radio tower that may also be used for cellular transmission

Digital Cell Phones

Cell phone technology took a huge leap forward with the introduction of digital phones. Analog and digital telephones are similar in that they transform sound waves from tower to tower. However, digital telephones transform sound waves into binary data while analog wave forms are transformed into electrical frequencies. Once digital sound waves have been converted they are then compressed. This compression allows between three and ten digital cell cells to occupy the space of a single analog call. Analog signals are not able to be compressed and manipulated as easily as a true digital signal.

Digital cell phones are required to contain a large amount of processing power. They use modulation and encoding schemes to convert analog information into digital. It is compressed and converted back while maintaining an acceptable level of voice quality. This process is often completed by using frequency-shift keying (FSK) which is sent back and forth over AMPS. FSK uses two frequencies, one for the binary 1s and the other for the binary 0s. FSK alternates rapidly between the two frequencies to send digital information between the tower and the phone.

In addition to AMPS, which is often used for analog technologies, two additional technologies are primarily used within the United States to access cellular airwaves: TDMA (Time Division Multiple Access) and CDMA (Code Division Multiple Access). The common consumer often will not need to be an expert on the differences between the technologies rather that they simply exist. Cellular phones can be identified as dual-mode or tri-mode. This label identifies the different ways a cell phone is able to access a cellular network. Having more access capabilities with your cell phone allows additional coverage options. Depending on where you are located geographically, you may have limited cellular coverage. If one access technology is not providing a strong signal for cell phone reception the phone will immediately change to using another access technology in hopes of finding a stronger signal.

Cell Phone Service Options

There are three popular cellular service options found within the United States. We have discussed AMPS which is used for normal analog cell phones. In addition to AMPS, digital cell phones and PCS are commonly uses. Digital cell phones and PCS are popular digital systems.

Digtial services uses digital phones but are found to communicate with normal AMPS towers. A call is initiated and uses a normal AMPS protocol. The conversion is then transmitted digitally. A digital phone is used to break a single AMPS voice channel into three digital channels. This technique is TDMA. This process allows three phones to share the same channel, each sharing time within that individual channel. Digital cell phones are a blend between the analog system and digital technology.

PCS phones use a separate set of towers and frequencies to separate themselves as completely digital. Their frequencies used are much higher which means their towers must be much closer. Encryption and error-correcting codes make the digital call much clearer and nearly impossible to intercept. PCS service is often able to include additional services into its cell package such as paging, caller ID or even email.

Summary

Cell phone technology took a huge leap forward with the introduction of digital phones. Digital cell phones are required to contain a large amount of processing power. They use modulation and encoding schemes to convert analog information into digital. Cellular phones can be identified as dual-mode or tri-mode. There are three popular cellular service options found within the United States. In addition to AMPS, digital cell phones and PCS are commonly uses.

Concept Reinforcement

1. Explain the link between telephones and radios.

2. Describe the three popular cellular service options found within the United States.

3. How are analog signals converted to digital?

Section 3.3 – Real-time Data Transmission

Section Objective

- Discuss real-time data transmission

Real-time data transmission refers to data that is available quickly enough after it is created that the time delay of the transmission goes unnoticed by the receiver. The convergence of telephony and the IT environment have worked on ideas together to converge since people started using digital voice coding. Transmitting data, voice and video applications using the same medium would make overall transmission paths much more efficient. Unfortunately, each of these applications has different needs.

Voice and video transmissions need a constant bandwidth and work to guarantee time of delivery. Data, on the other hand, does not need a constant bandwidth and does not even emphasize reliability of connection. Because of the differences networks have been structured differently to only meet the needs of the application they have been created for. For example, data networks allow everyone to use all available bandwidth to the maximum extent. Telephone networks reserve a channel per call. A channel is considered busy even if data is not being transferred (there is no talking by either caller).

VoIP

These differences and inconsistencies have led voice and video transmission technologies to use VoIP (Voice over IP). These technologies use VoIP over a real-time IP network (Internet) and have been developed as an alternative to the standard circuit-switching telephone network. VoIP is based on digital data transmission. VoIP only needs a connection to the Internet Service Provider to make a connection, allowing the user to skip any standard long distance changes they may otherwise occur using their standard phone line.

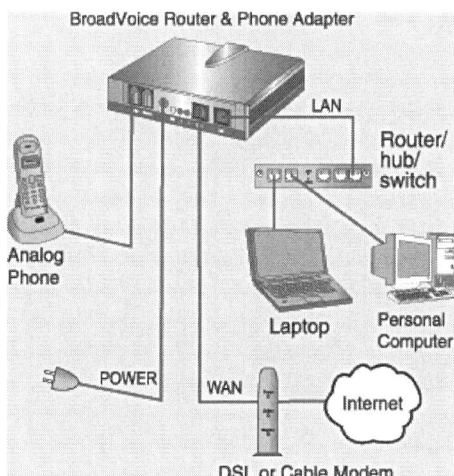

Common VoIP configuration

The first step in any VoIP call is completed with an Analog-to Digital Converter (ADC). The ADC works to convert the analog signal into digital data by dividing the signal into steps which are represented by numbers. The audio data is then compressed using a codec. This helps reduce the amount of digital data while maintaining audio quality. The compressed digital data is now able to be sent over the

Internet in layers (OSI Model) which ensure proper delivery.

The compressed digital data is formed into a data stream and divided into packets. These packets contain the audio data but also house information pertaining to the destination and the order in which the data should be encoded. Packets of information are sometimes dropped to ensure the overall quality of the information is maintained. Even though packets can be dropped there is usually enough information to make a phone conversation legible. The number of dropped packets will depend on the speed of your Internet connection as well as the distance between the two parties.

Dropped packets can be a benefit. If a small amount of information is delayed that will not impact the end result of a conversation it would be a benefit to simply drop the packet. The information can then be transmitted without having to wait on meaningless data. It can then be reassembled and will not have to impact the overall speed or quality of the final conversation. Once the voice data has arrived at its destination, it is reassembled in the correct order and converted back from digital to analog.

Summary

Real-time data transmission refers to data that is available quickly enough after it is created that the time delay of the transmission goes unnoticed by the receiver.

Transmitting data, voice and video applications using the same medium would make overall transmission paths much more efficient. Unfortunately, each of these applications has different needs. These differences and inconsistencies have led voice and video transmission technologies to use VoIP (Voice over IP). VoIP is based on digital data transmission.

Concept Reinforcement

1. Briefly explain real-time data transmission.

2. What does VoIP stand for and why was it originally developed?

3. Explain how data, voice and video applications work differently when transmitting information.

Section 3.4 – Delay Time and Data Rate

Section Objective

- Define and discuss delay time and finite data rate

Transmission Lines

Transmission lines are defined by delay time and data rate. Message delay is based on the information rate capacity. This tells you how long it takes voltage to travel from one end of the transmission medium to the other. The time delay is determined by the transmission speed as well as the length of the transmission medium. Data rate is determined not only by the transmission medium, but by the sender and receiver as well. Delay time and data rate combined have information engineers working to improve the speeds at which information can be both physically sent and received.

Wireless designs have become more prevalent over recent years. Digital designs have been called upon to reach higher frequencies to reach these wireless demands. The foundation of transmission speed is determined by delay time and data rate. Each of these can be measured and calculated within a transmission line. A **transmission line** is a set of conductors used for transmitting electrical signals. A conductor is a medium that conducts heat, light, sound, or electrical charge. Coaxial cables, twisted-wire pairs, and parallel-wire pairs are examples of transmission lines. These are each considered lossless transmission lines.

Delay Time

Transmission lines often encounter delay as **reflection** occurs. Whenever electrical resistance is encountered by a signal a portion of the signal is transmitted while some of the signal is reflected. Consider the following analogy: Imagine sitting in a small boat and looking down into the water. A fish swimming below the boat is able to see you from below the surface of the water because the water is transparent. In addition, you are able to see a portion of your image as a reflection upon the surface of the water. In this case, some of the light from your image travels through the water to the fish while some of the light reflects back to you. This same phenomenon occurs with electrical signal frequencies. Electrical resistance of a signal determines the amplitude of reflected and transmitted waves. As a signal travels down the medium, delay is associated with it.

Much like the fish reflection example, this picture shows a fish underwater
and a reflection of itself along the surface of the water.

Jitter is another potential cause for delay time within signal transmission. Jitter is a variation in delay time caused by queuing. Queuing is caused by congestion or route changes. Jitter is most likely to cause delay on slow or heavily congested mediums. These delay times are often found within VoIP when the transmission is competing with other processes within the same CPU. Jitter delays are often measured in milliseconds making the delay unnoticeable to the common user.

Data Rate

As programs and files are becoming larger, the highest possible data transfer rate is certainly becoming more and more desirable. **Data rate** defines the speed at which data can be transmitted from one device to another. Data rates are measured in megabits or megabytes per second. Data rates are determined by error expectation as well as transmission power. Data rates are necessary in order to increase accuracy by lessening the opportunity for error and transmitting information as efficiently as possible. There are typically trade-offs that occur with determining data rates. Speed must often be decreased to ensure accuracy and overall transmission costs are increased as increased power is required.

Summary

Transmission lines are defined by delay time and data rate. The time delay is determined by the transmission speed as well as the length of the transmission medium. Data rate is determined not only by the transmission medium, but by the sender and receiver as well.

Transmission lines often encounter delay as reflection occurs. Whenever electrical resistance is encountered by a signal a portion of the signal is transmitted while some of the signal is reflected. Jitter is another potential cause for delay time within signal transmission. Jitter is a variation in delay time caused by queuing. As programs and files are becoming larger, the highest possible data transfer rate is certainly becoming more and more desirable. Data rate defines the speed at which data can be transmitted from one device to another.

Concept Reinforcement

1. What are transmission lines used for?

2. Explain how reflection and jitter may lead to delay times.

3. Explain the importance of efficient data rate.

Section 3.5 – Bandlimits

Section Objective

- Describe bandlimits

Bandwidth

Bandlimits are often discussed within the context of bandwidth. Bandwidth plays a vital role in telecommunications. Its unit of measurement is the hertz. **Bandwidth** can be defined as the range between the lowest and highest frequencies used for a particular application. One of the most common definitions deals with signal power. Filters, amplifiers, transmission lines, and other electromagnetic equipment help to control signals in reference to the allowed bandwidth.

Bandlimit

The process of controlling or limiting the range of frequencies allowed to pass through a medium is called **bandlimit**. For example, audio often uses filters to control bandlimit sounds. An equalizer is a good visual example when considering bandlimits. An equalizer is electronic equipment that reduces frequency distortion. Audio sound can be chopped at either end of the audio wave prior to allowing the sound to pass through the medium. The passing audio remains intact but has been bandlimited.

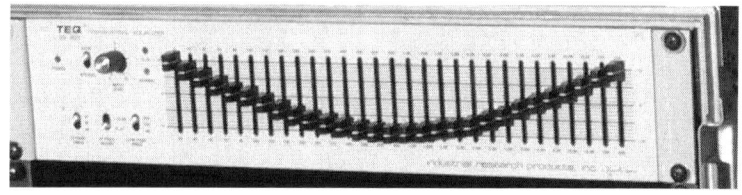

Audio equalizer

Radio stations as well as the FCC (U.S. Federal Communications Commission) have a very strict definition of bandwidth. Often it is referred to as the 20 dB rule or the 99 percent power containment rule. These rules pertain to the defined bandwidth allowed by individual radio stations. This bandwidth leaves exactly 0.5 percent of the signal power above the upper bandlimit and exactly 0.5 percent of the signal power below the bandlimit. This allows 99 percent of the signal power to occupy each band allowing for slight flexibility or error.

Cutoff frequency determines a bandlimited signal. If the amplitude is zero for all frequencies beyond an identified threshold then the remaining frequency will be removed, or cutoff. Prior to a frequency being cutoff it is often sampled.

Sampling Bandlimited Signals

Bandlimited signals have essentially been removed of their original information to allow the signal transmission to flow through the medium. Sampling allows a bandlimited signal to be closely reconstructed. As discussed with the Nyquist rate, a sampling rate that exceeds twice the maximum frequency in the bandlimited signal will allow the original signal to be reconstructed. The end result can be attributed to the Nyquist sampling theorem.

Summary

Bandlimits are often discussed within the context of bandwidth. Bandwidth can be defined as the range between the lowest and highest frequencies used for a particular application. The process of controlling or limiting the range of frequencies allowed to pass through a medium is called bandlimit. Cutoff frequency determines a bandlimited signal. If the amplitude is zero for all frequencies beyond an identified threshold then the remaining frequency will be removed, or cutoff. Prior to a frequency being cutoff it is often sampled.

Concept Reinforcement

1. Briefly explain bandwidth.

2. Explain how bandlimit and bandwidth are related.

3. What is the impact of a sampled bandlimited signal?

Section 3.6 – Wire Transmissions

Section Objective

- Explain wire transmissions

Proper wiring can effectively reduce noise interference and speed the delivery of your data from the originating transmission point. There are a number of considerations to make when choosing a medium for information transmission. Desired transfer speed, network designs, distance required as well as flexibility and ease of installation are each important factors in determining the proper transmission medium.

There are many different wiring options that are available to reduce unwanted noise pickup from entering the line. While noise and interference cannot be completely removed through wire transmission, good engineering and proper installation help to reduce the negative effects. Two main types of wires are found useful in wire transmissions including twisted pair and coaxial cable.

Twisted-Pair Wiring

Twisted-pair wiring is based on copper wire construction. It is a flexible cable that contains pairs of copper wires that are twisted together for reduction of interference. These wires are insulated with an outer jacket. Twisted-pair cables are often used for network setups between computers, usually in an office environment where connections are linked from computer to computer. Twisted-pair cables help in the elimination of noise due to electromagnetic fields by twisting the two copper wires at regular intervals. Errors or disturbance have the same magnitude based on this configuration and are able to be predicted resulting in error cancellation. While most twisted-pair cabling is shielded, unshielded cable is available. While still twisted together, there is no shielding material between the pair or outer cable jacket.

Example of wires within a twisted-pair cable

Coaxial Cable

A **coaxial cable** is another alternative for wire transmission that works to protect data from noise or interference. Coaxial cable, similar to twisted-pair, is also based on copper wire construction. Coaxial cable is a heavy-duty wire made up of an inner conductor which consists of several copper tubes used to carry a current. These tubes are separated and surrounded by an outer conductor shaped like a cylinder. The outer conductor is used to transmit electric impulses from one point to another. The entire cable is surrounded by a protective plastic casing. Coaxial cables are most often used to transmit telephone and cable television. Coaxial cables are ideal for wire transmission as they are not affected by external electric or magnetic fields which otherwise may cause interference in data transfer.

End connection of a coaxial cable

Difference between Twisted-Pair Wiring and Coaxial Cable

The difference between twisted-pair wiring and coaxial cable can be most easily identified through their name. Twisted-pair consists of a number of pairs of cables twisted together and used for a single signal. Coaxial cable uses an outer and inner conductor along the same axis, or alignment.

While both wire transmission options can be used in a similar setting, both have an advantage over another depending on your expectations. Coaxial cable follows a higher bandwidth and allows additional data to flow through the cable. Twisted-pair is more likely to encounter electrical interference. However, coaxial cable typically costs more than twisted-pair and is heavier and larger than twisted-pair.

Summary

Proper wiring can effectively reduce noise interference and speed the delivery of your data from the originating transmission point. Two main types of wires are found useful in wire transmissions including twisted pair and coaxial cable. Twisted-pair cables are often used for network setups between computers, usually in an office environment where connections are linked from computer to computer. Coaxial cables are most often used to transmit telephone and cable television. The difference between twisted-pair wiring and coaxial cable can be most easily identified through their name.

Concept Reinforcement

1. What benefits are found through proper wiring setups?

2. List four factors that must be considered in determining the proper transmission medium.

3. Describe the differences between twisted-pair wiring and coaxial cable.

Section 3.7 – Fiber-Optic Cable

Section Objective

- Discuss how fiber-optic cables are used to transmit data

Fiber-optic cables allow significantly greater transmission rates than any wire transmission. Fiber-optic networks are capable of carrying tens of thousands of telephone calls within a commercial network at one time. This equates to over 10 billion digital bits being transmitted per second. A fiber-optic system is similar to a copper wire system. However, fiber-optics uses light pulses to transmit information through hair-thin fiber lines instead of using electronic pulses to transmit information down copper wire lines.

Looking inside the core of a fiber-optic cable is similar to looking down a mirror that has been made to look like a long paper towel tube. Fiber-optic cables are composed of a core of hair-thin fibers. For very short distances the core can be plastic but most are made from glass which is used for longer transmission distances. A protective shell of material, the cladding, encloses the core. The cladding is typically covered by a plastic PVC outer jacket. Transmission occurs through fiber-optic cables as light rays are converted into digital pulses with a laser or light which moves along the core without damaging the protective shell of material.

> **Fiber Optic Advantages Over Copper Wire:**
>
> **Speed:** Operate at higher speeds
> **Bandwidth:** Larger carrying capacity
> **Distance:** Signals can be transmitted further without needing to be strengthened
> **Resistance:** Greater resistance to noise
> **Maintenance:** Cost much less to maintain

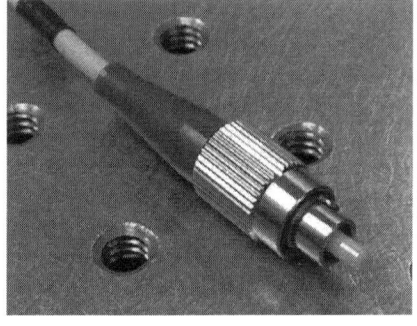

Fiber-optic wire

Signals are transmitted through fiber-optic cables in a similar path to wire. Fiber-optics is even used in conjunction with wire transmission mediums. At one end of the system is a transmitter. A transmitter is where the information, or signal, is started into the fiber-optic line. The transmitter is able to accept coded electronic pulses from a wire transmission. It will then translate and process that electronic information into coded light pulses. Light pulses are then funneled into the fiber-optic medium where they travel down the cable.

Light pulses stay within the core because of the internal reflection of light by the cladding. Light rays are reflected back to the core from the cladding in a similar fashion to a mirror-like tube. Think of a mirror shaped like a long paper towel roll. A flashlight can be held at one end of the tube and light would filter all the way to the other end because of the mirrors. Even if the tube were bent, the mirrors would reflect the light all the way through. This is how fiber optic cables work. Light pulses move easily down the fiber-optic lines because of a principle known as total internal reflection.

Fiber-optics became available in 1970 when Corning Glass Works was able to produce a fiber created out of glass. Fiber-optics was not feasible without the use of glass. Fiber-optics has since increased in popularity at a rapid pace. The first commercial installation was completed in 1977. Telephone companies have been gradually replacing their copper wire systems with fiber-optic lines and now use it as the backbone architecture to their long-distance connections between city phone systems.

Fiber-optic cable wrapped and hung on a telephone line

Cable television and power companies have also integrated fiber-optics into their cable systems by replacing outdated lines. Neighborhoods and homes have even begun to integrate fiber-optics into personal use. Fiber-optic and coaxial cable have been combined to form a hybrid system that allows individuals to benefit by using fiber-optic technology.

Groups of computers which are connected to each other for the purpose of sharing data is called a LAN (Local Area Network). Colleges, universities and office buildings are examples of primary users of LAN setups. These setups have also seen an increase in the use of fiber-optics to increase communication speed, traffic capacity and overall reliability.

Summary

Fiber-optic cables allow significantly greater transmission rates than any wire transmission. A fiber-optic system is similar to a copper wire system. However, fiber-optics uses light pulses to transmit information through hair-thin fiber lines instead of using electronic pulses to transmit information down copper wire lines. Transmission occurs through fiber-optic cables as light rays are converted into digital pulses with a laser or light which moves along the core without damaging the protective shell of material. Fiber-optics has since increased in popularity at a rapid pace. The first commercial installation was completed in 1977.

Concept Reinforcement

1. Describe the major benefit fiber-optic wiring has over copper wire.

2. How is data transferred through a fiber-optic cable?

3. List two major industries that now rely upon fiber-optic cable for faster, more reliable data transmission.

Section 3.8 – Radio Communications

Section Objective

- Describe radio communications system design

Modern day radio allows people to enjoy music, sporting events and featured news. It has become one of the most important means of communication throughout the 20ᵗʰ century. While one of the most impressive inventions of our time, the radio is fairly simple in how it works. Understanding how a radio works is essential in grasping the communication process. A radio contains a transmitter, receiver, capacitors, tuner, and a loud speaker.

> **Hertz (Hz) Conversions:**
>
> 1 kHz = 1,000 Hz
>
> 1 MHz = 1,000,000 Hz
>
> 1 MHz = 1,000 kHz

Radio messages can be anything from a public service announcement to a radio broadcast. The purpose of a transmitter is to take a message in the form of a wave and encode it. This encoded message is then decoded by the transceiver as it is received. Radios can burn themselves out if they do not have a way to store electrical energy. A capacitor is used to ensure that the exact amount of energy is being used within a radio. A tuner is used to ensure the frequency of an antenna is working correctly. A tuner allows you to determine between your AM and FM frequencies. Finally, a speaker is the device which converts an electrical signal into sound. Magnets are used within a speaker to create sound as electrical signals are applied to them.

How are Sound Waves Different from Radio Waves?

Sound waves and radio waves are often considered one in the same. Sound consists of pressure variations, or vibrations, in a medium such as air or water. Radio waves are electromagnetic waves just like light, infrared or X-rays. When a radio is turned on the sound can be heard because the transmitter at the radio station has converted the sound waves into electromagnetic waves. These waves are then encoded into a specific radio frequency range. Generally AM stations are found in a range of 500-1600 kHz while FM stations are found between 86-107 Mhz. Radio electromagnetic waves are used for radio broadcast because they can travel large distances through the atmosphere without fading.

Towers built at higher altitudes assist towards radio waves
traveling longer distances without fading.

Transfer Techniques

Radio stations transmit information in the form of waves by using three possible techniques; pulse modulation, amplitude modulation, or frequency modulation. Pulse modulation involves turning on and off the wave transmission. This technique uses a form of dots, dashes, and phrases and allows sound to be created and manipulated. Pulse modulation led to the creation of Morse code. Telegram communication services were created as a result of Morse code. A telegram is a written message intended to be delivered by telegraph which is a communications system that allows information to be transmitted over a wire through a series of electrical pulses.

Amplitude modulation (AM) transmits analog or digital data by varying the voltage of a current. Modulation is the altering of waves in order to transmit a data signal from one location to another. It is the oldest method of transmitting voice electronically. In AM radio, the amplitude of a fixed frequency (the station's channel) is modulated by the analog audio signal. Frequency modulation (FM) is also used in radio communication. It varies the frequency of the wave of a current in order to transmit analog or digital data. AM frequencies can be heard from up to 1,000 miles on a clear night. The disadvantage that AM frequencies have in comparison to FM frequencies is the poorer wave reception leading to a lower sound quality.

Radio communication has become a vital part of our society in a number of different ways. Public safety uses radio to help eliminate crime and catch criminals. Dispatchers are used to give and receive data through radio communication. Transportation, such as taxi cabs, depends on two way communication to receive instructions on where to go in order to pick up customers. The airline industry is also heavily impacted by the success of radio communication in order to keep the crew, passengers, and other airplanes safe.

Summary

A radio contains a transmitter, receiver, capacitors, tuner, and a loud speaker. Radio stations transmit information in the form of waves by using three possible techniques; pulse modulation, amplitude modulation, or frequency modulation. Pulse modulation involves turning on and off the wave transmission. Amplitude modulation (AM) transmits analog or digital data by varying the voltage of a current. Frequency modulation varies the frequency of the wave of a current in order to transmit analog or digital data.

Concept Reinforcement

1. What is the disadvantage that AM frequencies have in comparison to FM frequencies?

2. Explain how radio communication has become a vital part of our society.

3. What three techniques are used by radio stations to transmit information in the form of waves?

Section 3.9 – Satellite Communication

Section Objective

- Describe satellite communications systems

There are hundreds of satellites in operation used for a wide variety of purposes such as weather forecasting, television broadcast, radio communications as well as Internet communications. A **satellite** is a specialized wireless receiver and / or transmitter that is placed in orbit of the Earth by a rocket that has been launched into space. The primary role of a satellite is to reflect electronic signals. However, satellites are also used for observation. They are able to be equipped with cameras or sensors used to download images from the particular vantage point of the satellite.

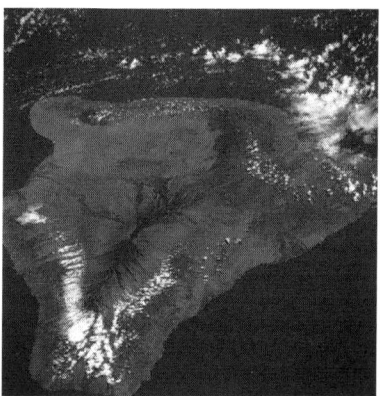

Satellite image of the Island of Hawaii

The first true satellite was called Telstar. Telstar was created by AT&T in 1962. Prior to Telstar, Russia launched an artificial satellite into space in the late 1950s. This satellite was about the size of a basketball and did nothing but transmit a simple Morse code signal over and over. Modern day satellite communication systems have been growing in sophistication and are now used to receive and re-transmit thousands of signals simultaneously.

Telstar, the first true satellite launched into orbit

Basic Elements

Satellite communications are comprised of the satellite itself as well as a ground station. The satellite is composed of three main sections including the fuel system, the controls and the transponder. The transponder includes an antenna, multiplexer, and frequency converter. The antenna is used to pick-up signals from the ground station. A multiplexer is used to allow the combination of more than one activity to occur within the satellite at one time while the frequency converter is used to reroute the received signals into the proper format.

The ground station is the portion of the satellite communication that is found on the Earth. The primary duty of the ground station is two-fold as it works with the uplink and downlink of transmitted data. In the case of an uplink, the satellite passes signals down towards the ground station which processes and converts the data as necessary. The downlink describes the opposite, data being transmitted from the ground station to the satellite.

Uses of Satellite Communication

Communication techniques can be accomplished in a number of different ways. As the distance between networks and importance of speed of transmission has increased, so too have the expectations of the communication pattern. From basic telephone networks to cellular communications and television signals, satellites are being used with more regularity and effectiveness than any alternative communication system.

Telephone networks have started to establish a satellite based link to connect geographically remote areas of the world. Countries with little or no telephone network infrastructure benefit from the effectiveness of a satellite connection. Cellular technology has seen bandwidth availability increased in an effort to expand the overall network connection possibilities through satellite connections. Satellite connections allow cellular users the availability of frequencies outside those of a ground based cellular network. Satellite communication has also been used since the 1960s to transmit broadcast television signals between television companies.

While the overall use of satellites is used by telephone and cable connections with increased regularity and success, satellite technologies are not without their limitations. Latency, poor bandwidth, and noise each hinder network connections and reliability.

Latency is considered a broadcast delay. Delay is more often noticed as high speed connections become faster. Due to the high altitudes of satellites in orbit, the time required for a transmission to navigate to and from a satellite can cause a series of small delays which eventually become noticeable. Poor bandwidth is often seen with the current limitations within the radio spectrum because there is a fixed amount of allocated satellite transmission available. Transmitted signals are known to encounter noise and ultimately a weak signal due to the distance between the original transmission and the satellite. However, these problems are often able to be solved by the proper error correction techniques.

Summary

A satellite is a specialized wireless receiver and / or transmitter that is placed in orbit of the Earth by a rocket that has been launched into space. The primary role of a satellite is to reflect electronic signals. The first true satellite was called Telstar. Satellite communications are comprised of the satellite itself as well as a ground station. The satellite is composed of three main sections including the fuel system, the controls and the transponder. The ground station is the portion of the satellite communication that is found on the Earth.

Concept Reinforcement

1. What is the name of the first true satellite launched into space?

2. A satellite is comprised of three main sections. List and explain each individual section.

3. What are the limitations of satellite technologies?

Section 3.10 – Global Positioning System

Section Objective

- Explain the Global Positioning System (GPS) and how it is used

Explorers and navigators have used a number of techniques to enable themselves to locate their position. The importance of a system designed to enable this to occur with any accuracy is important in trying to avoid tragedy or in reaching a final destination. In early times the use of stars in the sky were used by explorers to assist in navigation, with obvious negative limitations. However, on June 26, 1993, the U.S. Air Force launched the 24th NAVSTAR satellite into orbit to complete a network of 24 satellites known as Global Positioning System, or GPS. NAVSTAR stands for Navigation System with Timing And Ranging. This release launched the public usage of GPS. **GPS** allows you to instantly learn your specific location down to the exact latitude, longitude, or even altitude, within feet.

NAVSTAR GPS satellite undergoing pre-launch testing

Prior to the release of the 24th NAVSTAR satellite into orbit, GPS had been under development by the U.S. Department of Defense (DOD). The DOD launched the original group of 11 orbiting NAVSTAR satellites in 1978. These were used strictly for military use. The military use was needed to determine accurate positioning of submarines prior to launching missiles. Each of the old methods of determining position had their flaws. Atmospheric conditions, limited range or interference were each keeping the military from being as accurate and precise as needed. GPS technology was later released to the public when the DOD released the 24th NAVSTAR satellite into orbit. Each of these satellites have been placed in a precise orbit at an altitude of 10,900 miles. Each satellite weighs two tons, is 18.5 feet long, and orbits the earth in a little less than 12 hours.

GPS was originally made possible by the use of atomic clocks. Atomic clocks are the world's most accurate timepieces as they are able to precisely determine time within a billionth of a second. Atomic clocks were created by physicists simply seeking to answer questions about the nature of the universe. They had no concept that these clocks would someday lead to saving lives and generating thousands of jobs within a multi-billion-dollar industry that GPS has become. These atomic clocks combined with the developing GPS technology have enhanced the use of navigation, surveying, vehicle tracking as well as outdoor recreation such as hiking or camping.

Global Positioning System is a U.S. space-based radio navigation system. Along with the 24 satellites that orbit the Earth, GPS systems include five control and monitoring stations on Earth, and the GPS receivers owned by individual users. The combination of these three components results in reliable positioning, navigation and timing services on a worldwide basis. The control and monitoring stations are responsible for keeping the satellites in precise orbit. Each of the 24 satellites transmits their own unique signal.

Image of a GPS satellite in orbit

GPS is currently available to all, including civilians. A GPS receiver is the only piece of equipment needed for a civilian to incorporate the benefits of GPS into their everyday lives. The receiver is used to provide three-dimensional location (latitude, longitude, and altitude) as well as time. Each GPS receiver has 24 separate sets of data stored within them to identify the position of each of the 24 satellites. Each satellite has a distinct orbit which allows the receiver to be pinpointed within a few feet. This information is provided with accuracy through all weather conditions, day and night, anywhere in the world. The distance between each satellite and the individual GPS receivers is measured by the time it takes for radio waves to reach the GPS unit. Each GPS unit is required to have at least four channels while most now have more. Out of the 24 satellites orbiting the Earth, at least five are always accessible to any one point on Earth at any given point in time.

GPS can be used in any type of weather. It can be used on land, in the air, and for marine applications. However, GPS is not clear of all interference or limitations. Heavy tree cover or cliffs can be known to interfere with a GPS signal. Steep hills or tall buildings can also impede a quality connection. Many of these situations can simply be avoided by moving to a better location without getting too far off the intended route. From the very first explorer who asked the basic question, "Where am I?" to the technology used today with GPS, the benefits of GPS far outweigh the potential negatives.

Summary

GPS allows you to instantly learn your specific location down to the exact latitude, longitude, or even altitude, within feet. Along with the 24 satellites that orbit the Earth, GPS systems include five control and monitoring stations on Earth, and the GPS receivers owned by individual users. A GPS receiver is the only piece of equipment needed for a civilian to incorporate the benefits of GPS into their everyday lives. GPS can be used in any type of weather. It can be used on land, in the air, and for marine applications. Many situations that may cause interference with a GPS system can simply be avoided by moving to a better location without getting too far off the intended route.

Concept Reinforcement

1. Provide and example of how GPS was used by the military prior to being released to the public.

2. Describe the dimensions of a GPS satellite.

3. Explain the potential limitations that may cause interference with GPS signal reception.

Section 3.11 – Circuit-based Networks

Section Objective

- Discuss circuit-based networks

Many network operators find themselves maintaining an environment which contains a combination of different mediums, physical environments, transmission techniques and equipment all used together to accomplish the same end result. A number of functions are being migrated into the same network, often slowing down transition speeds and decreasing dependability. Along with the combination of different functions, networks have been upgraded, combined and connected as new, faster or better options emerge. While the goal is to improve transmission speed and accuracy, the combination of different networks can actually increase cost and make operational procedures very difficult to manage. Many of these consolidated services often include a combination of circuit based networks such as private voice transmission lines, frame relay and ATM systems. A circuit is an electronic path between two points. Circuit-based networks are one that obtains a connection between two end points within the network.

Circuit-based networks support a wide range of applications. Each circuit network provides different performance characteristics. Private line services provide support for voice traffic or high bandwidth applications that depend on delivery without delay or interference. Frame relay was designed to support higher levels of data traffic in the most efficient way. ATM was originally designed to support a variety of services that can support large amounts of data transmission or even act in much of the same fashion as a private line.

Private Line Services

Private line services are also known as circuit-switched. This is a circuit network configuration in which the physical path is obtained for a dedicated single connection. This connection is made between two end points in the network for the duration of the connection. These network connections are often dedicated to voice phone service which is known as a circuit-switched network. A specific path is reserved by the telephone company to the phone number you are calling. This path is maintained for the duration of your call, eliminating any other calls to take place within the same physical line.

Frame Relay

Frame relay is another circuit-based network. Frame relay is a network service designed for cost-efficient data transmission. It was designed to manage sporadic traffic between local area networks (LANs) and between end-points within a wide area network (WAN). It requires a dedicated connection during the transmission period. Because they require a steady flow of transmission, frame relay is not ideally used for voice or video. Error correction is left to the end-points of any data transmission through frame relay. It puts data in variable-size units called frames and drops any errors that are detected. This simply speeds up the overall data transmission.

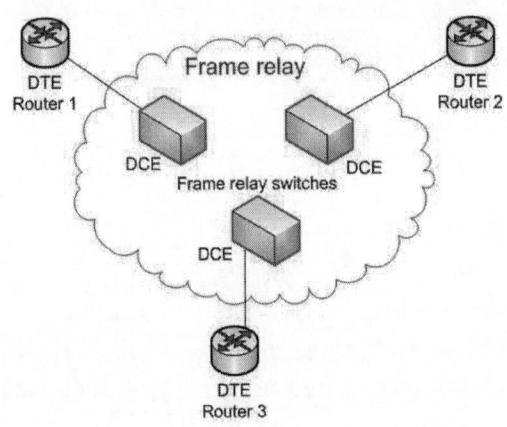

Example of a Frame Relay network setup

ATM

ATM (Asynchronous Transfer Mode) is a network technology based on transferring data in packets of a fixed size. Unlike private line services which only allow a dedicated single connection or frame relay which has been designed to handle specific type and amounts of data transmission, ATM allows for more flexibility and often greater capability. ATM transmits video, audio, and computer information over the same network. It keeps a balance between all transmissions allowing no single type of data to control an individual line. ATM creates a fixed route between two points whenever data transfer begins. Because of this design, ATM is less adaptable to sudden surges in network traffic.

Supporting Circuit-based Networks

Supporting circuit-based networks and services as they combine within each other becomes difficult. When circuit-based networks are used, data transmission is often able to transfer between separate designs and protocol. Performance, recoverability, availability and predictability become necessary in order to rely upon accurate and efficient information transfer. Networks must provide flexibility to meet specific requirements of each service supported. These networks must also be designed for failures. Consolidating network solutions becomes important when equipment fails. Each circuit-based service must also have its own operational environment for monitoring, troubleshooting and managing the network. When combined with thought and consideration, multiple services and environments can be supported. Operational costs can be limited and delivery expectations can be maintained when private lines, frame relay, and ATM are consolidated correctly.

Summary

A circuit is an electronic path between two points. Circuit-based networks are one that obtains a connection between two end points within the network. Circuit-based networks support a wide range of applications. Each circuit network provides different performance characteristics. Private line services provide support for voice traffic or high bandwidth applications that depend on delivery without delay or interference. Frame relay was designed to support higher levels of data traffic in the most efficient way. ATM was originally designed to support a variety of services that can support large amounts of data transmission or even act in much of the same fashion as a private line.

Concept Reinforcement

1. What type of information transmission are private line services often dedicated to?

2. What was frame relay designed to manage?

3. Why is ATM considered to have greater benefit than frame relay or a private line when transmitting data?

Section 3.12 – LANs

Section Objective

- Describe Local Area Networks (LAN)

What is a LAN?

Local area networks (LANs) differ based on the environment. LANs cannot be described except in the most general way because the technologies used in LANs are extremely diverse. Generally speaking, a LAN is a computer network ranging in size from a few computers in a single office to hundreds or even thousands of devices spread across several buildings. A computer network is a group of interconnected computers. They share a common communications line or wireless link. Resources within a LAN often function off a single processor or server within a small geographic area such as an office building.

Example of a LAN office environment

The first LAN was put into service in 1964 at the Livermore Laboratory. It was used to support atomic weapons research. LANs eventually spread to the public in the late 1970s. They were first used to create high-speed links between several large computers found at one site.

Access is shared within a LAN to printers, file servers, and other devices or services. A LAN provides a foundation to link computers together to allow this shared access. As few as two or three users may serve as a LAN. This may be the case in a home network. However, a LAN may also serve users well into the thousands.

Often we consider the geographic area of a LAN to include an area no larger than a building. However, similar network setups may be found within a college campus which is used to connect a number of buildings together. LANs may also be plugged into larger networks such as a WAN (Wide area network) or MAN (Metropolitan Area Network). A WAN is used to connect two or more LANs together so users at multiple sites are able to communicate with each other. LANs often include much higher data-transfer rates and do not have the need to lease telecommunications lines like a WAN setup would. A MAN (Metropolitan Area Network) is used to connect computers around a large city size area.

Generally speaking, larger areas covered within a LAN setup will typically have slower data transfer speeds. However, there are techniques and technology to help speed up the transfer rate and reliability of larger LAN setups such as protocols that transmit data across the network as well as software to interpret and negotiate the network. A protocol is a standard that controls or enables the connection, communication and data transfer between two computing endpoints. A protocol works to determine how best to share resources within a LAN environment.

Repeaters, bridges or routers also benefit towards faster, reliable information transfer. A **repeater** is a **network** device that is used to regenerate or replicate signals. A bridge filters data traffic at a **network** boundary and helps reduce the amount of traffic on a LAN by dividing it into two segments. A router is a computer networking device that forwards data packets across a network toward their destinations. The level of management required to run a LAN depends on the type, configuration, and number of devices involved.

Summary

A LAN (Local Area Network) is a computer network ranging in size from a few computers in a single office to hundreds or even thousands of devices spread across several buildings. A computer network is a group of interconnected computers. Access is shared within a LAN to printers, file servers, and other devices or services. LANs may also be plugged into larger networks such as a WAN (Wide area network) or MAN (Metropolitan Area Network).

Concept Reinforcement

1. When and where was the first LAN put into service?

2. Access is shared within a LAN to different devices. Name three devises other than a computer that may be connected within a LAN setup.

3. Discuss the difference between a Local Area Network and a Wide Area Network.

Section 3.13 – Ethernet

Section Objective

- Define Ethernet

Networking

Networking allows two computers to communicate. Information is able to be shared, sent and received through a network setup. Networks are used much more often than one may think. For example, the Internet is a network. It links millions of computers around the world. Public libraries have replaced many of the card catalogs with computer terminals that now allow you to search for a reference through the library network. Networks can also be found within an airport. Screens are spread throughout an airport to display information regarding arrival and departure information, all controlled through the airport network. In each example, networking allows shared information to be viewed through multiple devices setup in multiple locations.

The most common type of local area network is an Ethernet LAN. A **LAN** is a computer network ranging in size from a few computers in a single office to hundreds or even thousands of devices spread across several buildings. The smallest home LAN can have exactly two computers while a large LAN can accommodate many thousands of computers.

Ethernet

Many basic networking needs can be accomplished through an Ethernet LAN. Ethernet technology typically operates within a single building. Devices within an Ethernet setup are usually arranged in close proximity often with no more than a few hundred feet of cable between them. While Ethernet technology has increased its physical range, connections between geographically separated locations is often inefficient and impractical.

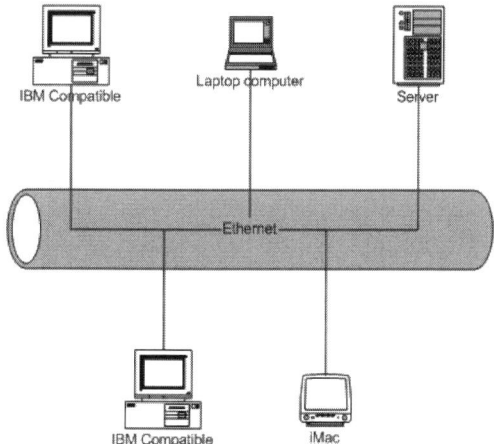

Example of an Ethernet LAN setup

Robert Metcalfe designed and tested the first Ethernet network in 1973 while working at Xerox Corporation's Palo Alto Research Center (PARC). The Ethernet concept originated from his development of linking a connection between a computer and printer. He worked on a standard to allow these two devices to communicate by using a physical cable connection. The Ethernet has since become the most popular and widely used network technology in the world because of its simple and logical setup potential. An Ethernet allows any device to communicate over a single cable once it is introduced to the network. This allows a small, simple network to expand to accommodate new devices without having to modify any devices already on the network.

Robert Metcalfe wearing the National Medal of Technology and speaking
in a ceremony at the U.S. Department of Commerce

As computer networking has expanded, Ethernet standards have grown. However, the basic foundation of an Ethernet is the same today as it was originally designed by Bob Metcalfe. An Ethernet setup is guided by protocol. Protocol refers to a set of rules that dictate and guide communication within an Ethernet. In order for two computers to successfully communicate they must both understand the same protocols.

Ethernets consist of a few components that are standard in the networking process. An Ethernet must first consist of two or more computers linked together, or networked. All computers must include a network interface card (NIC). A NIC card allows each computer to have a unique address so information can be guided to the correct destination. An Ethernet cable must also run from computer to a central switch or hub. The switch or hub identifies the NIC address as information is received. It then determines where the information needs to be sent and directs it across the LAN appropriately. This network setup creates a communication system that allows the sharing of data and resources, including printers, fax machines, and scanners.

Ethernet networks can also be wireless. Radio waves replace Ethernet cable as the foundation for two-way communication. Wireless NIC cards are used to identify individual addresses and small antennas are used in place of a switch or hub to direct information to the correct destination. These wireless Ethernet networks can be more flexible to use but may also require extra security considerations.

Summary

Networking allows two computers to communicate. Information is able to be shared, sent and received through a network setup. The most common type of local area network is an Ethernet LAN. Devices within an Ethernet setup are usually arranged in close proximity often with no more than a few hundred feet of cable between them. An Ethernet setup is guided by protocol. Protocol refers to a set of rules that dictate and guide communication within an Ethernet. In order for two computers to successfully communicate they must both understand the same protocols.

Concept Reinforcement

1. What is the most common type of LAN?

2. Who is credited with the design of the first Ethernet setup?

3. Explain what a NIC card does in an Ethernet setup.

Section 3.14 – Internet

Section Objective

- Discuss the organization of the Internet

Many people now use the Internet on a weekly, if not daily, basis. The Internet is relied upon for communication, news or weather updates, banking, photo sharing, as well as a number of other possibilities. The Internet is becoming so widely used that it is now relied upon from a personal, professional, and educational perspective. While many use the Internet on a daily basis do we really understand what it is? We often know what type of information can be found on the Internet but do we know how it gets there or how it is maintained?

The History of the Internet

The concept of the Internet began as ARPANET (Advanced Research Project Agency Network). ARPANET was a U.S. Department of Defense project to create a nationwide computer network. This network was designed to function even if a large portion of it were destroyed through nuclear war or a natural disaster. This project was funded by the U.S. Department of Defense in 1969. Through the next two decades the evolving network was used primarily by academic institutions, scientists, and government for research and communications. These groups appreciated and benefited from the ability to connect to other computing systems and databases as well as share information via E-mail. The Internet made a drastic change in 1992 when the U.S. government allowed commercial entities to offer Internet access to the general public for the first time. This change in focus led to a rapid growth in popularity and usability that has since changed the world we live in.

How Does the Internet Work?

The Internet is an organized combination of independent, interconnected networks. These networks support computer-to-computer communication through a standard group of protocols and procedures understood as the Internet Standards. The Internet is not individually owned or controlled by any one person or group. It lacks any central authority that controls the entire Internet. However, participation in the Internet requires you to follow a group of standards. Various governing boards work together to establish a group of standards that Internet users must follow. Internet providers adhere to these standards while also often making their network available to the public. It is the cooperation and understanding of the same group of standards followed by Internet providers that forms the Internet. Regardless of how an Internet connection is made, dial-up, high speed, or wireless, your computer connects to an Internet Service Provider (ISP). An ISP is a company that provides user access to internet services such as E-mail or the Web.

Common example of a satellite broadband Internet connection
made through a satellite dish outside your home.

Communication through an Internet Service Provider would not be possible without organized protocols. Protocol refers to a set of rules that dictate and guide communication. These protocols allow computers around the world to access and view specified files on other computers. TCP/IP is the most often used Internet protocol. It allows computers to describe data to one another over a network. TCP (Transmission Control Protocol) takes the information you want to send over the Internet and breaks it down into small pieces of data called packets. IP (Internet Protocol) makes sure the packets get to their final destination. Once the packets are delivered to their final destination they are reassembled by TCP into the original, recognizable information. TCP and IP combined allow information to be addressed, routed, and reassembled and are used each and every time the Internet is used.

Computer protocol is not limited to just TCP/IP. STMP (Simple Text Mail Protocol) works with E-mail. FTP (File Transfer Protocol) is often used for uploading and downloading files to and from other computers. HTTP is another protocol often noticed as you enter a Web address. Each of these protocols work together to ensure communication is able to occur between all computers through E-mail or Web pages.

What are the Benefits of the Internet?

The Internet allows a user to easily connect through an ordinary personal computer. Exchanging E-mail or posting information for other users to access is made possible through the Internet. Multimedia features such as music, television, movies, or photographic images can be uploaded, downloaded or shared via the Internet. Research and business data can be shared among colleagues. Problems or questions can be answered via the Internet as can valuable feedback or suggestions from customers, clients, or business partners. The Internet is also widely used for marketing and publicizing products and services. The greatest benefit of the Internet is that each of these diverse possibilities is not limited to simply one geographic location. Information can be researched, obtained or shared from anywhere around the world.

Summary

The Internet is relied upon for communication, news or weather updates, banking, photo sharing, as well as a number of other possibilities. The concept of the Internet began as ARPANET (Advanced Research Project Agency Network). ARPANET was a U.S. Department of Defense project to create a nationwide computer network. The Internet is an organized combination of independent, interconnected networks. Various governing boards work together to establish a group of standards that Internet users must follow.

Internet providers adhere to these standards while also often making their network available to the public. Communication through an Internet Service Provider would not be possible without organized protocols. Protocol refers to a set of rules that dictate and guide communication.

Concept Reinforcement

1. List five common uses of the Internet.

2. What does ARPANET stand for and what was its original design purpose?

3. TCP/IP is the most commonly used Internet protocol. Explain how TCP and IP work together to accomplish the overall goal of allowing computers to describe and share data to one another.

Section 3.15 – Information Security

Section Objective

- Explain the importance of information security

Why We Need Information Security

Both personally and professionally we hear of the importance of protecting your computer information. Identity theft has become one of the fastest growing crimes of our society as information technology has expanded. Security breaches occur on a professional level as companies have had personal information stolen from their computer systems. Millions of dollars are spent each year on protecting our computer and network systems yet millions of dollars in theft and damage recovery occur each year as computer crime continues to increase. Information security should be ranked high on the priority list of big corporate companies as well as the individual with one personal computer at their home.

Computer Security

Because of the wide variety jobs computers are relied upon to maintain, calculate, or store they often contain personal information of some sort. Credit card numbers, bank account information or personal identification information can all lead to a large loss in money and privacy. There are a number of steps that can be used to minimize the potential for loss and increase your expectation of computer security. One of the most important ideas to consider in working with computer technology is trust. If you cannot trust an Internet site you are viewing or an email that was sent to you do not go any further. Information security is often breached by intruders, or those who find their way into a computer system and do their damage from there. This can often occur by providing personal information through an Internet site that is not secure or accepting and opening and email from a sender you are not familiar with.

Steps to Take Towards Protecting Your Computer

One of the most important steps in securing all important computer information is to make a backup of all important files or folders. This holds true for both work and home environments. You should always consider which types of files are important enough to back up. If the file can be easily recreated or reinstalled and would not present any harm to your personal, professional, or financial history then the file probably does not need to be backed up. If you determine that a file should be backed up then you need to determine how you should do it and how often it should be saved. Portable storage devices are becoming more affordable and compatible with most computer systems. Saving and backing up files and folders as often as possible or certainly when additional important information is obtained will only save you from any potential future loss.

Important files and folders can be stored on an external hard drive
like this one from Iomega.

Many security attacks occur through email, attachments and downloads. Email viruses and worms can cause serious damage to a computer or specific files. A **computer virus** is a program that is designed to replicate itself by copying itself into the other programs stored in a **computer**. A **computer worm** is a program which copies itself across a network. A computer worm differs from a computer virus in that a computer worm can run itself. Viruses and worms are becoming so common that most that use a computer have either already experienced an attack or will at some point. Know where an email is sent from. Have you received an email from that sender before? Were you expecting to receive an attachment? Does anything about the email message look suspicious? Make sure you are familiar with the sender and certainly any attachments prior to opening them. If you are downloading an attachment or installing a program make sure you have a clear description of what the program does. This description can often be found on a web site where the download is from or from the sender of the message you received.

Another common mistake users make that could prevent information from being stolen or corrupted is to use strong passwords that are never repeated. Keeping up with all your passwords today can be difficult. However, without the proper care and consideration, a stolen password can lead to loss of important information or permit someone other than yourself to view your personal information or account numbers. Passwords should be complicated by making them as unique as possible. Using letters, upper and lower case, as well as numbers will strengthen your password. However, remembering your password without having to write it down is also important. You certainly do not want anyone without permission to obtain your list of personal passwords.

Finally, installing anti-virus programs, firewalls and encryption programs will greatly increase the security surrounding your computer. An anti-virus program looks at the contents of each file and searches for a specific pattern that matches a harmful profile. For each file that matches a potential threat an option is given to destroy the file or bypass the warning. A firewall acts as a guard to your computer. It looks at network traffic received from other computers. A firewall is important as it keeps all the unwanted out and permits only appropriate traffic to enter and leave your computer. File encryption programs transform information from one form to another. This transformation usually includes going from readable text to encrypted, or unorganized, text. Encrypted text remains unreadable until it is received by the appropriate location.

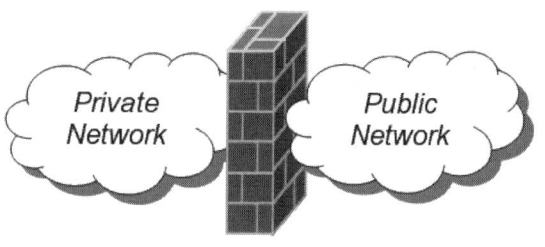

The brick wall depicts the job accomplished by a firewall.

Summary

Identity theft has become one of the fastest growing crimes of our society as information technology has expanded. Information security should be ranked high on the priority list of big corporate companies as well as the individual with one personal computer at their home. There are a number of steps that can be used to minimize the potential for loss and increase your expectation of computer security. You should always consider which types of files are important enough to back up and make sure they are maintained on a regular basis. Stolen password can lead to loss of important information or permit people other than yourself to view your personal information or account numbers making unique passwords more important. Installing anti-virus programs, firewalls and encryption programs will greatly increase the security surrounding your computer.

Concept Reinforcement

1. Discuss the different types of information that can be stolen or corrupted by not making information security a priority.

2. Why is it important to make all passwords unique?

3. Briefly describe the difference between an anti-virus program, firewall and encryption programs.

Appendix

Computer Engineering Answer Key – Unit 1

Section 1.1

1. 1969 became an important year towards the evolution of computers. Bell Labs developed their own operating system known as UNIX. ARPANet was launched which became one of the precursors to today's Internet. The personal computer (PC) was proposed for the first time by Alan Keys. Alan later became a designer for Apple. Intel was formed in 1969 after a group of technicians became unhappy with Fairchild Semiconductor and decided to form their own company.

2. ENIAC

3. Early in 1945, with the Mark I stopped for repairs, a moth was found inside a piece of the computer which resulted in technical problems. Fixing the system became known as "debugging" from that day forward.

Section 1.2

1. The central processing unit (CPU) can be known as the "brain" of the computer system. The CPU is often called a processor. It is a chip whose speed determines how fast the personal computer (PC) operates. Everything that a computer does is managed by the CPU. It performs commands and instructions while controlling the operation of the computer.

2. By referring to the clock within your computer system all components in a computer can synchronize properly.

3. Random-access memory (RAM) is used to temporarily store information which the computer is currently working. When you turn your computer off you will lose all data stored in RAM. Read-only memory (ROM) is a more permanent type of memory storage used by the computer for important data that does not change. Your computer is powered on and off each time by ROM.

4. monitor, keyboard, mouse, and removable storage

Section 1.3

1. Claude Shannon has been credited as the founder of information theory.

2. An encoder is a device used to change a signal, data, or information into a code. This code may serve a variety of purposes. It may compress information to be stored or encrypted. It may also be used to translate information from one code to another.

3. DSP (Digital Signal Processing) is a process of identifying signals by assigning a specific sequence of numbers or symbols as well as the processing of these symbols.

Section 1.4

1. Analog processes information in a continuous stream.

2. The word digital describes any system transmission based on discontinuous data or events.

3. modem

Section 1.5

1. The Base 10 number system represents 10 different digits that can be used to make any number we want. The numbers are 0,1,2,3,4,5,6,7,8 and 9. Each numerical position in the base 10 number system has a value that is ten times the value of the previous position. Computers use the base 2 number system. This means there are 2 different digits called bits that we can use to make any number we want rather than using digits.

2. The term bit means "binary digit", meaning 0 or 1 in binary numbering.

3. Microprocessors are programmed with and operate in binary code. The values from all inputs form sensors and switches that are converted to binary form before they are read by the microprocessor. All memory operations, all data communication, and all output commands are performed with binary code.

Section 1.6

1. The application layer is the OSI layer closest to the end user, which means that both the OSI application layer and the user interact directly with the software application. The presentation layer works to transform data into the form that the application layer can accept. The Session Layer controls the dialogues/connections (sessions) between computers. The Transport Layer provides transparent transfer of data between end users, providing reliable data transfer services to the upper layers. The Network Layer performs network routing functions. The Data Link Layer provides the functional and procedural ability to transfer data between network entities and to detect and possibly correct errors that may occur in the Physical Layer. The Physical Layer defines the electrical and physical specifications for devices.

2. Protocols ensure that computers everywhere can talk to one another. Protocols are used to determine the type of error checking to be used and data compression method, if any. They also determine how a device will indicate that it has sent a message and how the receiving device will indicate that the message has been received.

Section 1.7

1. File Transfer Protocol

2. The receiving of information is called downloading while the giving of information is called uploading.

3. Every operating system (OS) has its own way of organizing files, which are arranged in a way that makes it easiest for each OS to access data. Without FTP it would be difficult to communicate between two computers using different operating systems. For example, you could be working from a Windows operating system and the computer you want to retrieve files from is running Unix. Your Windows machine has no idea where to find the file you requested because it doesn't use the same logic as the Unix machine. FTP allows these different operating systems to communicate.

4. OPEN: This command initiates a connection between your computer (the client) and the other computer (the server) so that files may be exchanged.

    DIR: This lets your machine request a listing of the directories and their contents on the remote host.

    GET: This command requests that the file be transferred from the remote to the local host.

    SEND: This command works in reverse, delivering a file from your computer to a remote one.

    CLOSE: This ends the file transfer session.

Section 1.8

1. Transmission Control Protocol (TCP) is responsible for dissecting pieces of information so they can be transferred over the Internet and then reassembled.

2. Internet Protocol (IP) manages addresses and guides files to the intended destination.

3. TCP/IP is what carries out the basic operations of the Web. It is also used on many local area networks.

Section 1.9

1. When we think of saving information on or to our personal computer we typically think of devices such as USB sticks, zip drives, floppy disks or CDs. These are all considered types of storage, or secondary storage, where data can be saved. Secondary storage differs from primary storage in that it is not directly accessible by the CPU.

2. Off-line storage, also known as disconnected storage, is computer data storage on a device that is not under the control of a processing unit.

3. While backing up your computer will not save your computer from crashing, or keep viruses from being able to infect it, it will ensure that no matter what happens to your computer you will still have access to the files that are important to you.

Section 1.10

1. A vector image takes a digital image and gives it a two-dimensional or three-dimensional shape. Vector uses a sequence of commands or mathematical statements to create the digital image. A raster image represents an image in a series of bits of information which translate into pixels on the screen.

2. Data compression is the process of storing data in a format that requires less space than usual. Data compression is useful in communications because it enables devices to transmit or store the same amount of data in fewer bits.

3. JPEG is a widely used file format. It was created to store photographs. JPEG files are very efficient at reducing the file size. PNG is an acronym of Portable Network Graphics. This file format is intended for artificial images. These images have sharp edges or areas filled using a single-color. GIF is not used often for photographs or large artificial images. GIF is best used for small animated images or when needing a transparent background.

4. Image enhancement is the process of improving the quality of a digitally stored image by manipulating the image with software.

Section 1.11

1. A bitmap is a representation of the entire image consisting of rows and columns filled with dots. These dots allow the computer to form a graphic image. This image can then be stored within the computer memory.

2. The density of the dots is known as the resolution. The resolution determines how sharply the image is represented.

3. The value of each dot is stored in one or more bits of data. Each dot does not necessarily need to be filled. Neutral or colorless images may only need to be represented by one bit. However, colors and shades of gray require data consisting of more bits to better identify the color. The more bits used to represent a dot, the more colors and shades of gray that can be represented.

Section 1.12

1. Image synthesis is often referred to a rendering. It is the process of using a computer program to generate an image from a model.

2. Image synthesis usually involves a digital image or raster graphics image.

3. Digital rendering takes advantage of a number of features that may be altered in a variety of ways in creating a final product.

Section 1.13

1. Hard drive, printing images, CDs, DVDs

2. A memory card is a method used to store your images as you shoot with your digital camera.

3. The primary difference is storage space. A DVD will allow much more storage than a CD which may significantly cut down on the amount of discs you need to store and organize.

Section 1.14

1. Virtual Reality Modeling Language

2. box, cone, cylinder and sphere

3. VRML provides a variety of basic functions which are designed to run on all platforms. This has limited VRML capabilities from becoming the best all around virtual reality system. Other systems have been created for specific tasks or configured to run on specialized hardware. Limitations on VRML lighting and color have made it hard to overcome some of the more sophisticated, specialized virtual reality systems that continue to be developed.

Section 1.15

1. A scene graph is the structure of the world being created. The VRML file format allows you to create a scene graph using words and punctuation.

2. Nodes, which are part of a scene graph, describe a variety of items. Nodes may include a major feature, like a physical object in the world or even color of a specific object.

3. Individual nodes are capable of fitting together in a logical order. The node above is called the parent node and the node below is called the children node.

Computer Engineering Answer Key – Unit 2

Section 2.1

1. Information compression is useful because it helps reduce the use of expensive resources and increases the speed at which you send and receive information. It can cut down on your hard disk and transmission space by minimizing the information being sent or saved.

2. Claude Shannon has been credited with creating the first fundamental papers on information theory in the late 1940s and early 1950s. Shannon's definition of information revolves around identifying redundancy.

3. Lossless compression is a technique used to eliminate redundant information. Since only redundant information is removed it is often possible through lossless compression to restore the original information. Lossy compression is used when some loss of fidelity is acceptable in reducing the amount of data as much as possible.

Section 2.2

1. Probability coding is based on the ability to compress or simplify data by using probabilities.

2. Probability coding is only effective if there is a good idea of what kind of files are most likely to be transmitted.

3. As a lossless compression technique, one file shrinks as another expands.

Section 2.3

1. Variable length coding maps symbols to a variable number of bits. Variable length coding is a form of lossless data compression because it is able to encode and decode with zero errors.

2. David A. Huffman

3. Huffman coding is rather fast, but does not produce an efficient compression ratio. Arithmetic coding uses a coding table that is updated frequently to reflect real time distribution statistics.

4. Lempel-Ziv coding is one of the most popular compression techniques for lossless data storage.

Section 2.4

1. Universal coding techniques have the capability to globally compress all types of data without loss.

2. Universal coding is typically more complex than other compression technologies and has a slower overall speed in which data is transferred.

3. Universal coding works well when the probabilities of particular variables are unknown. It spots sequences that have previously appeared and uses this repeated information to build a more efficient message.

Section 2.5

1. Image compression is a technique used to make the file size of an image smaller.

2. It becomes acceptable if the image quality is not greatly impacted while the file size is significantly decreased.

3. GIF

Section 2.6

1. It breaks down video information into small packages of data. An analog signal is sampled. The video packages are analyzed at various points in time. The difference between the actual sample value and its predicted value is quantified, or specifically defined, and then encoded. This process allows a digital value to be formed.

2. When editing, copying, or working complex special effects, the 4:2:2 will reveal a much greater quality advantage.

3. Sampling rate is the rate at which analog signals are converted into digital form while sampling size is the number of times a signal is sampled in order to accomplish the proper conversion.

Section 2.7

1. The major advantage of MPEG compared to other video and audio coding formats is that MPEG files are much smaller yet still allow similar overall quality.

2. The International Telecommunications Union (ITU) and the International Standards Organization (ISO) worked together to standardize digital video compression.

3. The MPEG-1 compression standard is primarily used for moving pictures and audio. Digital television and DVD compression is based on the MPEG-2 compression standard. It was developed from MPEG-1, but designed for the compression and transmission of digital broadcast television. MPEG-4 has become the standard for multimedia and Web compression.

Section 2.8

1. DTV is more efficient and flexible than the traditional broadcast technology known as analog. It enables broadcasters to offer television with crystal clear pictures and CD quality sound. Interactive video and data services are also possibilities that DTV may offer over analog technology.

2. HDTV is the highest quality of DTV but is a separate format. It offers the best available picture resolution, clarity and color. It also provides theatre surround-sound and a wide screen format which gives a 'movie-like' format to the viewing experience. SDTV, standard definition television, is another common format. Having DTV does not automatically allow a consumer to view a television picture in high definition. Specific television sets are needed to receive high definition television programming.

3. Analog television sets, Digital-ready sets, HDTV-ready sets

Section 2.9

1. Low frequencies are known as infrasound while the high frequencies are known as ultrasound.

2. Movement is measured by an energy which occurs within a medium such as air, water, or any liquid or solid matter. Sound waves are heard as a source is disturbed or vibrates. This can be a ringing telephone or a voice of a person. These sounds disturb a particle in the surrounding medium and those particles disturb those next to them and so on as they travel through the medium. This disturbance creates movement in a wave pattern much like waves of water. The wave then carries the sound energy through the medium until it is received by the source.

3. Isaac Newton developed a mathematical theory of sound in 1687 as documented within his book, *Principia*. He believed that sound could be interpreted as pressure pulses that are transmitted through side-by-side particles.

Section 2.10

1. Signals can be expressed through a number of different mediums such as electronic voltage, magnetic particles, radio frequency waves, or even pulses of light.

2. If sound falls outside the frequency range it becomes unusable.

3. PCM is based on a sampling rate needed to convert an analog signal to a digital signal. Sampling is the process of replacing portions of the analog signal with amplitudes taken at a regular interval. These amplitudes are measured and given a valid value. The sampling rate allows the original sound to be created again when necessary.

Section 2.11

1. hertz (Hz) or cycles per second (cps)

2. Selecting a medium bandwidth large enough to fit the audio signal is vital towards successful audio signal transmission. This often requires a compression technique to allow the overall bandwidth to be manageable.

3. In measuring the frequency of audio signals, electromagnetic waves, or signals, are measured in number of cycles per second. The overall range of these radio or light frequencies, or the difference between the highest and lowest frequency measured, is known as bandwidth.

Section 2.12

1. Sampling allows the audio signal to be reconstructed to its originally state successfully

2. The Nyquist Sampling Theorem states that the sampling frequency must be greater than twice the bandwidth of the input signal. This allows the signal to be reconstructed to its originally state successfully.

3. The job of the analog-to-digital converter (ADC) is to store the value of the signal as a digital number. Once the signal is sampled by the ADC, a digital to analog converter (DAC) can be used to reverse the process. The DAC uses the digital signal to reconstruct the sound back into an analog form.

Section 2.13

1. Digitizing

2. The human ear is able to interpret audio samples by picking up vibrating pressure waves.

3. In order to properly digitize audio samples you must have the proper equipment. An audio source such as a CD is needed to house the original digital source. A computer will also be needed to house the transferred sound from the audio source. Transferring sound from an audio source to a computer is done via an audio cable. Software will be helpful in allowing the user to digitize and edit sounds once stored onto the computer. Headphones or speakers are helpful towards allowing the sound to be adjusted or edited.

Section 2.14

1. Quantization works along with compression to lessen the overall size of files.

2. They are both motivated by the need to reduce the amount of data needed to represent a signal.

3. When a JPEG image is encoded, the data representing the image is processed and then quantized and coded. The number of bits needed to represent the image can reduce the precision of the transformed values.

Section 2.15

1. It can be found as an individual component within a stereo system. A CD player can also be a handheld, portable device. CD players can also be self-contained units that include an amplifier and speakers.

2. A CD is 4.72 inches in diameter, 0.047 inches thick, and weighs about 0.53 of an ounce.

3. The compact disc standard was adopted in 1981 and the first CDs and CD players were introduced in the market in 1982.

Computer Engineering Answer Key – Unit 3

Section 3.1

1. Improvements and progress in technology and communication fields have led towards the shift from using electronic pulses (analog) to using bits of data (digital) to transmit sounds through a telephone. Prior to the change in primary telephone communication, analog telephones were the original telephone technology.

2. Scottish inventor, Alexander Graham Bell, received the first patent and was credited with the invention of the telephone at age 29 in 1876.

3. NAMPS is a version of AMPS. NAMPS differs from AMPS in that it incorporates some digital technology to allow the system to carry about three times as many calls as the original AMPS version.

Section 3.2

1. Cellular technology, analog or digital, is often referred to as telephones. However, they are more accurately referenced as a highly sophisticated radio.

2. AMPS is used for normal analog cell phones. In addition to AMPS, digital cell phones and PCS are commonly uses. Digital cell phones and PCS are popular digital systems.

3. Modulation and encoding schemes are used to convert analog information into digital.

Section 3.3

1. Real-time data transmission refers to data that is available quickly enough after it is created that the time delay of the transmission goes unnoticed by the receiver.

2. Voice over Internet Protocol has been developed as an alternative to the standard circuit-switching telephone network.

3. Voice and video transmissions need a constant bandwidth and work to guarantee time of delivery. Data, on the other hand, does not need a constant bandwidth and does not even emphasize reliability of connection.

Section 3.4

1. A transmission line is a set of conductors used for transmitting electrical signals.

2. Whenever electrical resistance is encountered by a signal a portion of the signal is transmitted while some of the signal is reflected. Jitter is a variation in delay time caused by queuing. Queuing is caused by congestion or route changes.

3. As programs and files are becoming larger, the highest possible data transfer rate is certainly becoming more and more desirable.

Section 3.5

1. Bandwidth can be defined as the range between the lowest and highest frequencies used for a particular application.

2. Bandlimits are often discussed within the context of bandwidth. The process of controlling or limiting the range of frequencies allowed to pass through a medium is called bandlimit.

3. Sampling allows a bandlimited signal to be closely reconstructed.

Section 3.6

1. Proper wiring can effectively reduce noise interference and speed the delivery of your data from the originating transmission point.

2. Desired transfer speed, network designs, distance required as well as flexibility and ease of installation are each important factors in determining the proper transmission medium.

3. Coaxial cable follows a higher bandwidth and allows additional data to flow through the cable. Twisted-pair is more likely to encounter electrical interference. However, coaxial cable typically costs more than twisted-pair and is heavier and larger than twisted-pair.

Section 3.7

1. Fiber-optic cables allow significantly greater transmission rates than any wire transmission.

2. Fiber-optics uses light pulses to transmit information through hair-thin fiber lines instead of using electronic pulses to transmit information down copper wire lines.

3. Cable television and power companies

Section 3.8

1. The disadvantage that AM frequencies have in comparison to FM frequencies is the poorer wave reception leading to a lower sound quality.

2. Radio communication has become a vital part of our society in a number of different ways. Public safety uses radio to help eliminate crime and catch criminals. Dispatchers are used to give and receive data through radio communication. Transportation, such as taxi cabs, depends on two way communication to receive instructions on where to go in order to pick up customers. The airline industry is also heavily impacted by the success of radio communication in order to keep the crew, passengers, and other airplanes safe.

3. Radio stations transmit information in the form of waves by using three possible techniques; pulse modulation, amplitude modulation, or frequency modulation.

Section 3.9

1. The first true satellite was called Telstar.

2. The satellite is composed of three main sections including the fuel system, the controls and the transponder. The transponder includes an antenna, multiplexer, and frequency converter. The antenna is used to pick-up signals from the ground station. A multiplexer is used to allow the combination of more than one activity to occur within the satellite at one time while the frequency converter is used to reroute the received signals into the proper format.

3. Latency, poor bandwidth, and noise each hinder network connections and reliability.

Section 3.10

1. The military use was needed to determine accurate positioning of submarines prior to launching missiles.

2. Each satellite weighs two tons, is 18.5 feet long, and orbits the earth in a little less than 12 hours.

3. Heavy tree cover or cliffs can be known to interfere with a GPS signal. Steep hills or tall buildings can also impede a quality connection.

Section 3.11

1. These network connections are often dedicated to voice phone service which is known as a circuit-switched network.

2. It was designed to manage sporadic traffic between local area networks (LANs) and between end-points within a wide area network (WAN).

3. Unlike private line services which only allow a dedicated single connection or frame relay which has been designed to handle specific type and amounts of data transmission, ATM allows for more flexibility and often greater capability. ATM transmits video, audio, and computer information over the same network.

Section 3.12

1. The first LAN was put into service in 1964 at the Livermore Laboratory.

2. Printer, fax machine, file server

3. A WAN is used to connect two or more LANs together so users at multiple sites are able to communicate with each other. LANs often include much higher data-transfer rates and do not have the need to lease telecommunications lines like a WAN setup would.

Section 3.13

1. The most common type of local area network is an Ethernet LAN.

2. Robert Metcalfe designed and tested the first Ethernet network in 1973 while working at Xerox Corporation's Palo Alto Research Center (PARC).

3. A NIC card allows each computer to have a unique address so information can be guided to the correct destination.

Section 3.14

1. The Internet is relied upon for communication, news or weather updates, banking, photo sharing, emailing, file sharing, etc.

2. ARPANET (Advanced Research Project Agency Network) was designed to function even if a large portion of it were destroyed through nuclear war or a natural disaster.

3. TCP and IP combined allow information to be addressed, routed, and reassembled and are used each and every time the Internet is used.

Section 3.15

1. Credit card numbers, bank account information or personal identification information can all lead to a large loss in money and privacy.

2. Without the proper care and consideration, a stolen password can lead to loss of important information or permit someone other than yourself to view your personal information or account numbers.

3. An anti-virus program looks at the contents of each file and searches for a specific pattern that matches a harmful profile. A firewall acts as a guard to your computer. It looks at network traffic received from other computers. Encrypted text remains unreadable until it is received by the appropriate location.

8250390R00089

Printed in Great Britain
by Amazon.co.uk, Ltd.,
Marston Gate.